AUTO ENGINES OF TOMORROW

Power Alternatives for Cars to Come

Auto Engines of Tomorrow

POWER ALTERNATIVES

FOR CARS TO COME

Harris Edward Dark

INDIANA UNIVERSITY PRESS

Bloomington & London

Published in Canada by Fitzhenry & Whiteside Limited,
Don Mills, Ontario
Manufactured in the United States of America

Library of Congress Cataloging in Publication Data
Dark, Harris Edward, 1922–
    Auto engines of tomorrow.

    Bibliography
    Includes index.
    1. Automobiles—Motors. I. Title.
TL210.D33   1975   629.2'5   74-6518
ISBN 0-253-10490-4   75 76 77 78 79 1 2 3 4 5

To Phyl again, for more hard work, with love

Contents

Acknowledgments

For help in the compilation and checking of the facts, figures, and events covered by this book, I owe many thanks to the public relations and engineering departments of the Motor Vehicle Manufacturers Association of the United States; Chrysler Corporation; North American Automotive Operations of Ford Motor Company; General Motors Corporation; American Motors Corporation; Curtiss-Wright Corporation; the Johnson Division and the Evinrude Division of Outboard Marine Corporation; Toyo Kogyo Company, Ltd.; Mazda Motors of America, Inc.; Dr. Felix Wankel; Chevrolet Motor Division of GM; Oldsmobile Division of GM; Detroit Diesel Allison Division of GM; water engineer Irwin Brown; The American Society of Mechanical Engineers; Chek-Chart Corporation; Professor Lloyd Orr, Indiana University; Lear Motors Corporation; Thermo Electron Corporation; Aerojet Liquid Rocket Company; Scientific Energy Systems Corporation; Pancoastal, Inc.; Eaton Corporation; Dean David V. Ragone, College of Engineering, University of Michigan; Allis-Chalmers; Electric Fuel Propulsion Corporation; Charles T. Zegers, project manager of Electric Vehicle Council; Volkswagen of America; Peugeot, Inc.; John J. Brogan, director, and George M. Thur, branch chief, Alternative Automotive Power Systems Division, Environmental Protection Agency; ESB Incorporated; Mercedes-Benz of North America, Inc.; Elmer Whiddon, Texas Tourist Development Agency; and the Beaumont, Texas, Chamber of Commerce.

AUTO ENGINES OF TOMORROW

Power Alternatives for Cars to Come

Introduction

It all depends. . . .

That sounds like a pop song title and it probably is. Pop or not, it's the theme song of the seventies for Detroit, where decisions affecting future automotive engine design and production have never been so difficult.

America, indeed the world, has been struggling with an air pollution crisis for over a decade. The pollution has been around much longer than that—its effects, "big-city air," have been increasingly noticeable for at least four decades. The automobile, an early suspect, was proved in the fifties to be a smog producer, and the subsequent government regulations to combat auto exhaust emissions forced the automotive industry to develop techniques and devices for emissions reduction.

Those measures, which included detuning the conventional engine, reducing its compression ratio, and in many cases after-treating its exhaust, had a generally debilitating effect on the engine that spoiled its fuel conomy. The emissions engineering, incidentally, consisted entirely of *modifications* to the existing conventional engine. The only new engine introduced was the Wankel, and it, after all, is also an internal-combustion (rotary) piston engine that has the same emissions problems as all the others now on the road, with one redeeming quality: its small size allows more room under the hood for emissions-control equipment.

Introduction

During 1973, the year in which the national fleet topped the 125-million-vehicle mark, including more than 101 million passenger cars that were using more gasoline-per-mile than ever before, we started running out of gasoline and diesel oil at the retail level. In spite of brimful storage tanks, in 1973 we saw lines of cars at filling stations, truckers threatening nationwide strikes because of the diesel fuel situation, and *economy* becoming a word to be spoken with reverence. When, in late 1973, the Arabs cut off supplies of Middle East oil, the shortage problem compounded and a nation panicked at the thought of losing its wheels.

Whether the fuel shortage of 1973 was genuine or whether it was contrived by those who reaped enormous profits from it may never be known. Politicians, even those not discredited by Watergate and its backlash, seemed reluctant to investigate a monumental industry like petroleum. With its retail price doubled, gasoline was flowing freely in 1974, but there was still a fuel crisis: a fuel-*price* crisis that seemed even more threatening than the air pollution crisis. The two crises together brought tremendous pressure to bear on the auto industry to take corrective action in two different directions.

Of course the automotive industry has a long history of experimentation with various kinds of engines, and in annual reports, news releases, and public appearances of company officials, that point is frequently and emphatically made. This may be why the average person finds it incredible that all the money and time spent by the major industries known collectively as "Detroit" cannot turn up a "different" engine that is superior in all respects to the *one* gasoline-fueled internal combustion automobile engine the industry has used from its beginning.

Remember, the industry has been selling cars in record-breaking quantities since World War II and has billions of dollars tied up in tools for making the conventional engine. There hasn't been much motivation for bringing something different to the market, assuming it were possible to do so.

The pollution and fuel crises are beginning to provide motivation for serious and expensive research and development aimed at an engine that will be an improvement over the conventional one. Perhaps the most serious threat to the conventional engine is the 0.40 grams-per-mile maximum NOx requirement of the Environmental Protection Agency's emissions standards for 1977. It will be extremely difficult to meet this standard with a conventional engine, and the word around Detroit is that if the requirement is not relaxed, other types of engines will soon have a chance to compete with the conventional engine of today, but if the standard is relaxed or postponed, emergence of new engines will likewise be delayed.

Keep in mind that the apparent "uncertainty" in Detroit is not that at all. The decision makers are not paralyzed by ignorance, although because of proprietary considerations Detroit often appears uninformed and unprepared. Detroit pays a lot of money, hundreds of millions per annum, to keep itself fully informed on which developments are worthy of consideration and which are not. What is Detroit doing right now? Keeping itself in position to move in any desirable direction and in several different directions at once, if necessary.

This book is an attempt to show, in brief overview, some of the major steps and mis-steps that have resulted in today's conventional engine, some of the activity now being devoted to finding tomorrow's engines, and some of the opinions and other indicators that point to the probable winners in this biggest of all automobile races.

Could Detroit build a nearly pollution-free engine? Certainly. One that runs on less fuel? Indeed. When? When it must.

1

Piston, Pollution, and Petroleum

FOR MORE than thirty centuries man dreamed of a substitute for his beasts of burden and his own muscles while his role was little better than that of an ox, an ass, or a horse. Without realizing it he was longing for the Industrial Revolution.

That massive substitution of mechanical power for animal and human strength is said to have taken place in some forty-five years during the latter half of the eighteenth century and the beginning of the nineteenth century. In a sense, the Industrial Revolution is still going on and may never cease. But certainly those forty-five years saw more mechanical inventions than had occurred in all the preceding history of the world. The production of materials was rather suddenly moved from the home shop to the factory, where a relatively few workers could use large, expensive, and power-consuming machines to "manufacture" (make by hand) a volume of goods that was enormous in comparison with what their forefathers could make, literally, with their own hands only.

The object of all the achievements in early industrial manufacturing was the application of *power* to the job at hand. Such power, eventually to reach a magnitude equal to the muscle strength of thousands of humans or animals, had to come from an outside source—outside the factory, that is— and had to be available in large quantity. Windmills were well known, but in England, where the Industrial Revolution started, such devices were not considered satisfactory for powering factories. Water power was used a great deal at first, until the steam engine was perfected for factory work.

Steam, as an element for providing power through expansion, was known to Hero of Alexandria two centuries before Christ. But like the first automobiles, early steam machines were thought of as toys to be played with purely for amusement. For some reason it did not occur to any scientist until the eighteenth century that a steam engine could be put to a practical use. One explanation for steam's slow development undoubtedly lies in the fact that it was extremely dangerous to work with in the days before metallurgy had provided plumbing of integrity. All persons had respect for the damage potential of boiling water, and it's a safe bet that untold numbers of experimenters were injured by steam explosions down through the ages.

The first device that could be properly called a steam engine (that is, a train of mechanism forced by the pressure of steam into action that performs work) was invented by English partners Thomas Newcomen, Thomas Savery, and John Calley and patented in 1705. James Watt won fame and fortune years later by inventing so many valuable improvements that he often is mistakenly called the father of the steam engine. The first practical and successful steamboat was not built until about 1785; it was a product of John Fitch in the United States. And George Stephenson, a Newcastle, England, engineer who built locomotive engines from 1814 to 1833, is considered the first successful producer in that line. In 1830 the first transatlantic steamship voyages took place.

A French engineer, Nicolas Joseph Cugnot, built a self-propelled vehicle in 1769. By that description, the Cugnot machine was probably the first automobile; it was steam-powered, with *external* combustion. That setup involved a furnace in which a fuel (originally wood, later coal, coke, charcoal, and finally oil) was burned in the vicinity of a boiler

5

containing water. The resulting steam was piped to an engine with pistons, rods, and shaft capable of converting the steam pressure into rotary motion that could be utilized for doing work.

In subsequent chapters it will be shown that steam can be a working fluid in a modern machine of considerable efficiency, causing some serious engineers to believe that the steam engine will be a prominent automobile powerplant in our future. It is interesting to note at this point that the steam engine was the first practical engine, and that the automobile with steam power preceded the locomotive and the ocean liner by more than a generation.

The early steam engines and the road vehicles (called steam buses) they powered were extremely heavy and bulky. They were unsuited to the few good stagecoach roads, and they were obnoxious because of the noise and smoke they emitted, frightening man and beast alike. Moreover, a few well-publicized explosions that resulted in fatalities gave them a poor public image.

When it began to appear that in spite of such difficulties the steam bus might enjoy some success in the English transportation business, stagecoach owners and investors in railroads joined forces to quash the upstart vehicle. In 1865 the notorious Red Flag Law was rammed through Parliament. That incredible act limited the maximum speed of self-propelled, road-using vehicles to three miles per hour and additionally prohibited them from operating at all unless each was preceded along the road by a man carrying a red flag. Since the steam bus at that time was the only self-propelled road vehicle, it was virtually put out of business.

And so were most other efforts involved in the English development of the automobile. Few inventors could become interested in a vehicle that was required by law to hold its maximum speed to 3 mph. So while automobile development proceeded at an ever quickening pace in France, Germany, and the United States, it languished in England for thirty years, until the legislative roadblocks were repealed in 1896. By that time the putt-putt of the *internal*-combustion engine was being heard in all developed nations, and the electric car had reached such a high degree of development that in 1896 a Riker Electric won the first American automobile track race, at Narragansett Park, Rhode Island. Its average speed was 26.8 mph.

At this point, steam—which had accumulated by far the greatest number of years of development—and battery electricity and internal combustion were the only contestants in the race for automotive supremacy. They were running neck and neck in popularity, and at the turn of the century the outcome was still anybody's guess. Before another decade had passed, however, the issue was decided in favor of spark-fired gasoline.

The reasons are numerous and logical. Internal combustion from the very beginning was neater, easier to fit into a small package, and minus all the plumbing—with an explosion potential at every joint—that steam required. The electric car, on the other hand, could not carry enough fuel, in the form of batteries of electrochemical cells, to propel it more than a few miles, and never at satisfactory speeds. (True, the Riker Electric won its first race, but never again was any electric car considered a speed competitor of either gasoline or steam.)

The "invention" of the automobile was actually a development involving many people over the course of nearly a century. The engine principles of the steam bus of 1801 were further developed and refined for locomotives and ocean liners, but it was three more decades, in the 1860s, before *internal* combustion reached a point practical enough to be considered for automobile application.

In 1825 Michael Faraday extracted benzene, a type of gasoline, from tar derived from coal. It was quickly accepted by scientists and became one of the first liquid fuels capable of being used successfully in internal-combustion engines. However, it was not used as such a fuel for many years.

Auto Engines of Tomorrow

A Belgian-born Frenchman, Etienne Lenoir, put together a small internal-combustion engine in 1860. It was fueled by illuminating gas (gasoline-like distillates and spirits were available but not yet thought of as engine fuel), developed 1.5 hp (equal to the smallest modern power mower), and had a top rotating speed of 100 rpm (a modern V-8 cruises at about 3,000 rpm). The car in which the original Lenoir engine was installed could travel, nonstop, about six miles in an hour and a half.

Just a year later Siegfried Marcus, a German-born Austrian, achieved a remarkable improvement in Lenoir's concept. Marcus fashioned a two-stroke-cycle engine (of a type seen today in motorcycles, outboards, some farm equipment, and some power mowers) with a carburetor that mixed liquid fuel and air and with an electric-spark ignition similar to that used by many cars today. Once this had been accomplished (in 1861), many other inventors and investors became interested in the idea of an auto that could carry a tankful of petroleum fuel and perhaps travel hundreds of miles without refueling.

Of course that was all hypothetical because there were few roads capable of providing means of real travel. A whole economy of taxes and road building had to develop first. Bear in mind also that the first automobiles were built like wagons or buggies, animal-drawn vehicles designed for speeds of under 10 mph. They had wooden spoke wheels with iron tires. Their springs, if any, were primitive; there were no shock absorbers. Thus, hitting a bump at, say, 15 or 20 mph could be more than merely uncomfortable—it could be dangerous. Not even the foolhardy considered autos safe. Adequate brakes would have helped. So would headlights, horn, and a dependable steering mechanism, but it would be many, many years before these were accomplished facts.

At the turn into the twentieth century the three competitors—steam, electricity, and gasoline—were fairly even in prominence. At that point, historians and engineers have been known to agree, certain sets of circumstances could have allowed or forced the contest to go in any of the three directions.

For example, if the steam engine could have been simplified and made smaller, lighter, cheaper, and safer, it's possible that most of us would be driving autos today that are powered by steam; and if the steam engine existed in the market to a significant degree, the gasoline engine might be found only in power mowers, outboard boats, snowmobiles, and portable powerplants for electricity generation.

On the other hand, if someone at the turn of the century had come up with an electric battery of cells that could take a fast charge (say, in ten minutes) to prepare the battery for fueling an auto trip of a hundred to two-hundred miles, it's not unimaginable that we would have electric cars today that could travel at 100 mph for several hundred miles. If auto development had taken that direction, we would live in a quite different world today. For one thing, the demand for electric power would have been so great as far back as the 1920s that we would today have many more hydroelectric dams in operation and many more coal- and oil-burning electricity-generating plants to supplement the hydros.

Had this been the case, our air would now be cleaner, for two reasons: the hydros would be more numerous, and they would be producing zero pollution; and the oil- and coal-burning plants, although polluters to some degree, would be polluting air out in the country or in suburban areas. Today's automobiles, however, pollute in direct proportion to their population, which is more congested in the downtown areas. As we reach the three-quarter mark of the twentieth century, our congested areas are greatly troubled with pollution, and much of the downtown pollution emanates from the tailpipes of automobiles.

At the beginning of 1975 there were about fifty nuclear power plants in operation, with definite plans afoot to construct about ten times that many within the next five years. If today we had electric cars in numbers comparable to the prevalence of gasoline-burning vehicles, the urgent need for vast increases of electricity undoubtedly would have promoted the construction of nuclear power plants in much greater numbers than we have experienced so far, and that would have resulted in cleaner air throughout the country, indeed the world.

But it didn't work out that way. With the steam car loaded with plumbing and the electric car woefully short of power, the lighter, simpler "gas buggy" at the turn of the century began slowly, then rapidly, to take over the market. Two accidental happenings early in the twentieth century did more, it is said, for the development of the "gas buggy" than any other historical event. A new oil well and a fortunate factory fire were to provide such impetus as to insure dominance of the gasoline engine to the present day.

Early in 1901 the fabulous gusher Spindletop came in near Beaumont, Texas. Spindletop produced oil in such monumental quantities that that well's output alone dropped the price of crude to below five cents a barrel (less than one one-hundredth the going price today), and gasoline was so cheap to produce that the retail cost of motor fuel added practically nothing to the financial burden of owning and operating a motor vehicle. Even with the taxes that were quickly added for

Spindletop oil field, near Beaumont, Texas, in 1902. At this time there were a reported 138 producing wells here, and it was said that a person could walk the field from end to end on derrick floors without touching the ground. (COURTESY OF THE BEAUMONT, TEXAS, CHAMBER OF COMMERCE)

the purpose of building streets and highways, the low-cost availability of gasoline was a powerful incentive to build engines that would burn it as fuel.

The second and completely unrelated incident that established the dominance of the gasoline-piston engine happened a few months later: A fire completely destroyed the Detroit automobile factory of Ransom Eli Olds, one of the pioneers whose name still appears among brands being manufactured today. Faced with the prospect of being competitively forced out of business by the delay involved in building a new factory, Olds inadvertently embarked upon a system now known as mass production.

The only thing salvaged from the big fire was a complete prototype of the upcoming model of Olds' famous "curved dash" automobile. Taking it apart piece by piece, Olds jobbed out the manufacture of the car's components to machine shops all over Detroit. When parts were copied and produced in

Ransom E. Olds poses in his famous curved-dash Oldsmobile during a spin on a cold, snowy day near his home. (COURTESY OF OLDSMOBILE DIVISION, GENERAL MOTORS CORP.)

quantity, the pieces were delivered to Olds, who merely assembled them into cars. This meant that parts had to be interchangeable, and that the quantity of cars assembled depended only upon the supply of parts available. This method allowed Olds to produce and sell 425 cars in 1901 and more than 2,000 cars the following year.

As a direct result of Olds' innovation, many small machine shops in the Detroit area found themselves in the automotive business (where they elected to stay) and Detroit became the Motor City. Most important of all, the "gas buggy" went into mass production, and no type of engine other than the gasoline-fueled, spark-fired, reciprocating-piston engine has been important since in the automotive field.

Of all the engines considered by this book, the piston engine is one of the oldest, most complicated, and most familiar. The actual workings of the piston engine will require the least amount of explanation because of its ubiquitousness and consequent familiarity to the majority of today's drivers. Yet it is necessary to describe some of its aspects so that later they can be compared with those of possible future competitors.

The piston engine is so called because it contains pistons, which are placed inside cylinders with the closest possible fit consistent with lubricated movement against friction. The pistons are joined to a common crankshaft via connecting rods that are spaced evenly, in degrees that add up to a full circle of 360, so that, for example, in a 6-cylinder engine the journals where the rods are connected to the crank are spaced 60 degrees apart. The most popular internal-combustion engine is the four-stroke-cycle type, with four episodes for each cylinder occurring every two revolutions of the crankshaft.

As the engine revolves, the pistons travel up and down (or back and forth) in the cylinders in what is called a *reciprocating* action. This action, incidentally, is the main design difference between the conventional engine considered in this chapter and the Wankel rotary engine, which is the subject of

Chapter Five. The Wankel has pistons that do not follow a reciprocating motion but instead travel a circular path always in the same direction, hence "rotary." Many steam engines have reciprocating pistons; others are turbines; but all steam engines are *ex*ternal-combustion machines, that is, their fuel is always burned on the outside of the engines themselves. In the past few years it has become customary to pin the "piston" tag on the conventional reciprocating-piston, internal-combustion (fire on the inside) engine, the engine used in more than 99 percent of the autos now in existence.

In the four-stroke-cycle engine, there is usually one exhaust valve and one intake valve for each cylinder. The valves are operated by a camshaft that is geared to the crankshaft for invariable timing. The four strokes involved in a complete cycle occur in this order: intake, compression, power or combustion, exhaust. Looking for the moment at only one cylinder, we find that at the beginning of the intake stroke the intake valve is open, and as the piston pulls away from the intake valve a partial vacuum, a suction, is formed. This pulls an air-gasoline mixture from the carburetor, through the intake manifold, and into the combustion chamber of that one cylinder. When the piston reaches the end of the intake stroke, the intake valve closes. Then with both the intake and exhaust valve closed the piston reverses its direction and begins its compression stroke. Near or at the end of this stroke, the ignition system supplies a spark to the spark plug in this combustion chamber only. The piston reaches the end of its compression stroke and the fuel charge is under maximum compression, having been reduced in volume to about one-eighth its original size, and under several hundred pounds of pressure (depending on engine design).

When the spark occurs the fuel charge is flashed and begins a fast but even burning process that is *not* an explosion in the sense of a detonation. As the flame front moves across the combustion chamber the fuel charge is changed to a rapidly expanding gas that pushes the piston down on its power stroke. This is the only stroke of the four in the cycle that provides power to the crankshaft. At the end of this stroke the camshaft pushes open the exhaust valve and with a loud "pop" the still-burning and expanding gases blow themselves into the exhaust system, to be quieted by the muffler and, in late model cars, to be treated for emissions reduction. With the exhaust valve open, the piston completes the cycle by moving toward the exhaust valve, thus clearing the cylinder of most of the combustion products from the last firing. When the exhaust stroke is completed, the exhaust valve closes, the intake valve quickly opens, and the piston begins its intake stroke, the first of the next cycle.

Each of the other pistons in the engine has in turn followed the same continuous four-stroke pattern, so that in a running engine the episodes follow each other quite rapidly. Each cycle is completed in two engine revolutions, so that for each revolution in a 6-cylinder engine, three cycles involving twelve episodes will occur. Therefore, even when the engine is idling at, say, 500 rpm, six-thousand episodes occur every minute. The rapidity of occurrences helps make the engine run smoothly and, within certain practical limits, the faster the engine runs, the smoother it should be. This is particularly noticeable when an engine is defective in some minor way, such as an ignition misfire in one or two cylinders: the engine will limp and vibrate badly at idle but tend to smooth out at faster speeds, because the duration of the missing power strokes is shorter.

The conventional automobile engine must have a fuel system composed of a tank, line, pump, filter, and carburetor and an intake manifold, which is an integral part of the engine itself. The engine must also have an ignition system of battery, coil, distributor, spark plugs, and various switches, wires, and cables (often called "accessories"). It must have a liquid lubrication system of oil pan containing several quarts of oil, a pump and delivery system covering all the internal friction points, plus a filter. Then there must be a cooling system of radiator, pump, about ten to twenty quarts of coolant, two or more hoses, and numerous channels cast into the water jackets (chambers carrying coolant from the radiator) of the engine block and head(s). Internally, the crankshaft drives the camshaft to operate the valves, the distributor, and the oil pump.

The generator and the water pump are necessities driven outside the engine by belts from the crankshaft pulley. There may also be pulleys for a power-steering pump, air conditioning, and, in certain large engines, an air pump to supply fresh air to the exhaust manifold for antipollution exhaust treatment.

Dozens of different kinds of successful internal-combustion gasoline engines have been invented for auto applications. After a century of development, the conventional automobile engine has become extremely versatile, and versions of it have been put to every possible automotive use. For this reason alone it will be difficult to replace it economically in the marketplace even if a drawing-board replacement of merit is brought forth in the near future.

What was there about the gasoline reciprocating engine, what magic did it possess? How did it happen to "take hold" with the public and dominate the twentieth century's production of automobiles, small buses, and light trucks?

Probably the key word connected with the dominance of the reciprocating piston engine is *convenience*. Early automobile buffs put up with a lot of problems, but it always seemed that any problem not solved was on the verge of being solved, and the consuming public had a strong faith that the auto makers were sincerely trying to make a better and better product every year. This was generally true.

Early in the car-selling game, it was discovered that one of the major deterrents to purchase of autos for and by women was the physical problem of difficult cranking, especially in cold weather. In any season, cranking was a touchy and dangerous undertaking. It took a considerable amount of brute strength and not a little skill to turn over any engine, let alone a reluctant one. And if by chance the spark were advanced too far, so that a firing spark plug ignited a fuel charge ahead of a compressing piston, the engine would be spun rapidly backwards, jerking the crank from the cranker's hand, and perhaps spinning the crank around to strike the cranker on the wrist, often with bone-shattering force.

From the beginning, women loved automobiles and were resentful of the physical limitations they suffered in the operation of cars, especially in the cranking, which was obviously primary to going anywhere. Car makers were concerned too—they knew they could sell many more cars if women could be accommodated. Charles Kettering, then a young engineer for Cadillac who would spend a long life improving automobiles, was especially concerned with the cranking problem, and he resolved to do something about it. The electric motor seemed to be the only possibility for whirling the engine into motion in the same way that people had to crank it.

Kettering consulted top electrical people, one of them said to be Westinghouse himself, about the problem, and they all gave him the same answer: an electric motor strong enough to crank an engine would be outrageously large and would require enormous batteries to put it into motion under load. Undeterred by the opinions of experts, Kettering thought over the matter privately and at leisure. One day it occurred to him that all the electrical experts he consulted had been thinking about the problem in the same frame: with nothing in mind but a continuous-duty motor, one that could be operated all day without overheating. That was not what Kettering needed. He needed a motor that was capable of taking in a large amount of energy from a battery *for a short time* and using that energy to whirl the engine for only a few seconds until the engine started as if a person were cranking it. The type of starter motor that Kettering designed, therefore, would burn up if operated for several minutes instead of several seconds;

but that didn't matter—it was needed for only a few seconds. That type of motor, incidentally, is what every modern car now uses, with only a few minor improvements.

As to questions about the so-called magic of the gasoline reciprocating engine and how it achieved its dominance, one point of view has been that the public had no choice, that the manufacturers provided only the one type of engine (the gasoline-fired piston) and the public was stuck with it. What is probably closer to the truth is that everyone liked the piston engine because it did everything anyone required.

It certainly provided quick starting (once it was properly cranked), easy and fast warm-up, and—above all else—quick and powerful acceleration. And as automotive requirements increased, many fine engineers added to the growing fund of knowledge about the piston engine, and it was in a state of constant improvement for more than fifty years. The advancements covered a wide range of physical design, metallurgy, components, and accessories; and they included also the vast strides in lubricants and fuels made by the petroleum industry.

To suit the desires of the public (desires that sometimes had been cleverly "sold" to the public), automotive designers provided ever bigger and more powerful cars. With an abundance of gasoline to feed the behemoths, the public charged hither and yon in what is now remembered as an excess of enjoyment of the large machines. There was no publicity about impending shortages of fuel and other materials needed to make cars and keep them going. The United States was brimming with products and it was almost a virtue to do as we were bid by the advertisers and to waste things that in the middle 1970s we have come to regard as precious.

But long before there was any talk of shortages, another problem developed in the United States and other nations around the world. Suspect for many years, the automobile was proved in the early 1950s to be a major contributor to photochemical smog. That phenomenon is the result of complex changes in an atmosphere that has first been instilled with the gaseous products of combustion and then exposed to the rays of the sun for extended periods of time.

Los Angeles is now world-famous for its production of photochemical smog, although almost every people-congested area in the world has suffered from it during the past decade. It just happens that Los Angeles is the perfect laboratory for putting together three important elements in the development of that special scourge of modern mankind: a rich concentration of exhaust products from factories, oil refineries, and internal-combustion engines by the millions; a basin area in which the atmosphere frequently becomes trapped for several days; and a low-latitude radiation from the sun that readily converts the gaseous chemicals to nightmare effects that are hostile to humans, animals, and plants.

When, in the 1950s, it became evident that the automobile engine had the potential of poisoning vast numbers of people, animals, and plant life if its exhaust were allowed to go unrestricted, remedial action slowly got under way. Private persons, then consumer advocates, then politicians, then government agencies, then (note this) last of all, automobile manufacturers began to think and talk about ways of solving the problem of exhaust and other types of engine emissions.

An easy one was handled by the automotive industry first: crankcase emissions. Actually exhaust gases gone wrong, crankcase emissions are combustion products that "blow by" the pistons, are delivered to the crankcase by combustion pressure, and on older cars are forced to the outside by their own volume and by the blower action of the crankshaft as it revolves in the operating engine.

The disposition of blowby gases was once a simple matter: they were merely vented to the outside via what was known as a "road draft tube," something that looked like a short tailpipe. The gases were something that drivers put up with until it became common knowledge that as much as 25 percent of automotive pollution could emanate from this source. Early in the 1960s, the American automotive industry moved to correct the problem of blowby gases.

The technique, now called positive crankcase ventilation (PCV), simply called for eliminating the old openings to the crankcase (in the cap of the oil-filler tube and the road draft tube itself) and then providing a fresh-air input from the air filter via a hose or pipe. Also provided was a hose that pulled gases from the crankcase and delivered them to the intake manifold, so that they were burned with the regular fuel.

The system, which works well on new engines, has two flaws that cause trouble as cars grow older: The suction side must be controlled by a special valve called the PCV valve, which is subject to jamming due to gum deposits, causing the entire PCV system to become inoperative. Thus, the PCV valve must be checked and perhaps replaced every few thousand miles. And when an engine becomes worn, the amount of blowby gases can overwhelm the PCV system, causing throwing of crankcase oil and other problems.

The PCV emissions-control system was installed in many new cars in the early 1960s. While on new cars it virtually eliminated pollution from crankcase blowby gases, it was not designed to have any effect on pollutants emitted by the exhaust system. No important moves were made in that direction until the 1968 models were designed. At that time, ignition and carburetion changes of a minor nature were instituted and continued through the 1969 and 1970 models.

The 1971 models represented a drastic change in engine design, including a severe degradation of efficiency, horsepower, and fuel economy, primarily as a result of spark retardation and other ignition system modifications, and a reduction of compression ratio amounting to about 20 percent. The 1972, 1973, 1974, and 1975 models showed continuing loss of efficiency that caused certain models of cars, foreign as well as domestic, to deliver as little as 50 percent of the miles-per-gallon delivered by comparable automobiles of the 1967 model year.

Auto Engines of Tomorrow

During the eight-year period 1967–1975, the total of the nation's vehicles continued to increase dramatically, often by as many as a million units a year, and during the model years 1971–1975 about fifty million new cars were built, many of which used twice as much gasoline as models built in the middle 1960s. The main reason for the enormous increase in gasoline consumption was the government-mandated use of emissions-control equipment, adjustments, and modifications.

Whereas the piston engine as we knew it up until the late 1960s was more than enough to satisfy our needs and desires, suddenly it found itself on the side of the devils. First, we found it guilty of fouling our air. Then we discovered that to clean up its exhaust we had to practically ruin it. In the 1970s, the piston engine became hard to start, hard to keep running, slow to warm up, cantankerous in traffic, and often next to impossible to turn off. We learned that these ill-running engines were using two gallons of gas to one for a ten-year-old counterpart, that that problem threatened to run the whole nation out of gas, and that, with a shortage created for them, the oil companies could and did raise the price of gasoline to double what we paid just a few years earlier.

By the mid-1970s the piston engine had become more and more unsatisfactory, and its degradation was going to continue as more and more government restrictions on exhaust emissions were forced onto the motor vehicle manufacturers. When the public discovered that the majority of 1975 models thrust upon it were equipped with catalytic converters, indignation rose to new heights. The "cat" converters are muffler-like devices installed in addition to a muffler for the purpose of causing chemical changes in the exhaust gases to make the emissions less poisonous to humans, animals, and plants. The actual catalyst in each converter is a noble metal such as platinum or palladium, making the new devices cost from $50 to $150.

A major problem with catalytic conversion is that autos

so equipped must never be fed gasoline containing lead or additives such as ethylene dibromide used with lead in gasoline. Although by the fall of 1974 gasoline without additives harmful to "cat" converters was supposed to be universally available in order to fuel the 1975 models, it was folly for anyone to believe such a revolution could occur in the time allotted, and the change was indeed sluggish in taking place. Even tiny traces of certain additives in the fuel can permanently poison a "cat" converter. This means that trucks, tanks, plumbing, and pumps must be free of contamination from prior delivery of old-fashioned gas and must therefore be brand new in most cases. Drivers in metropolitan areas had little trouble finding "unleaded" gas (although much of it was being pumped from old equipment previously used to handle leaded gas and was presumably contaminated). In isolated areas, however, purchase of pure unleaded gasoline remains a hit-or-miss proposition; and in Canada, according to late-1974 warnings from the American Automobile Association, obtaining unleaded gasoline varied from easy to impossible. An owner of a poisoned "cat" converter must either drive a polluting car or pay for replacement of the catalyst or buy a whole new converter.

Like a bumbling politician who finds himself at odds not only with the opposition but also with members of his own party, the gasoline-piston engine found itself in real trouble in the middle 1970s. Those who have opposed the cars for years now find their forces augmented by former lovers of the conventional automobile. For the first time, really, we see serious efforts to discover or uncover a replacement for the gasoline-fueled, spark-fired, reciprocating (piston), internal-combustion engine. Strides have been made in several directions, in some cases resulting in more progress than was expected. But in most cases so far, the results have been disappointing. Americans have become accustomed to stopping wars with atomic bombs, to building the tallest buildings, the longest bridges, and the highest dams. We have frequently put men on the moon. It is hard for us to believe that we can't come up with an engine that is nonpolluting yet economical to operate.

In the following chapters the candidates that are possible replacements for the piston engine will be explored. Some of them are worthy of high hopes; others are little more than pipe dreams. But of the group considered in this book it is likely that one or two, or possibly more, will emerge as practical replacements of the conventional automobile engine that we know today. It is not *likely*, but it is possible, that a complete replacement will take place in the twentieth century. Between now and the turn of the new century we will see many changes in what we have considered the standard or conventional automobile. Many of the changes will be under the hood, which is where the remainder of this book will largely be spent.

2

Old Steam and the New Steamers

NICOLAS Joseph Cugnot, an eighteenth-century captain of French artillery, must have had unusual ability with mechanical things considering the limited progress that had been made up to his time with mechanisms utilizing steam power. After studying the possibilities of steam locomotion, Cugnot designed and constructed in 1769 a steam vehicle that could carry four passengers at an average speed of 2.25 mph. Cugnot's vehicle is generally accepted as being the first wheeled machine capable of self-propulsion from a source of on-board fuel—mankind's first automobile by technical definition.

With the idea that his new invention should be capable of towing heavy artillery pieces, he obtained permission from his minister of war, the Marquis de Choiseul, to conduct experiments. Cugnot's original car no longer exists, but one of the results of the early experiments does. It's a large, heavy, self-propelled carriage on display today in the Conservatoire des Arts et Métiers in Paris. Construction of the vehicle's engine was made possible by a device invented by the French general Gribeauval and built a short time before for machining cannon bores. Gribeauval's device facilitated the fabricating of a precisely bored cylinder, something that had been extremely difficult. The first Cugnot was a huge tricycle-chassis arrangement entirely built of wood overhung at the front by a large double-walled boiler allowing space between the inner and outer vessels for a fire grate. A copper tube connected the boiler to two vertical cylinders, which received steam under pressure. Pistons slid inside the cylinders; they were connected through two rods and two cranks to the single front wheel that provided both power and steering to the vehicle. There was a transmission, but its plans were lost to subsequent generations; it probably was a rack-and-pinion type, with a circular gear turning against a flat, toothed bar —the same type of gearing arrangement used today for steering mechanisms. The vehicle was about twenty feet long and eight feet tall at its highest point, with rear wheels approximately six feet in diameter.

The first trials, held at Vincennes, France, in the presence of a number of high-ranking officials, were successful. The large vehicle could haul five tons and travel at about the speed of a man walking slowly. Some time after its initial successes, the Cugnot ran into a wall, demolishing the wall and turning itself over. This erratic action of a machine that seemed to have a mind of its own frightened the minister of war, who, facing the economic crisis that preceded the French Revolution, became discouraged and withdrew the funds necessary for Cugnot to continue his experiments. Although Cugnot died a pauper in Paris in 1804, he left a heritage that may make it possible for all of us to enjoy the advantages of steam cars by the year 2000. Already, most of the formerly expensive components have been reduced in price and improved in quality so that, in today's steam cars, we see vehicles of considerable promise.

While the steam car (or truck, wagon, or tractor, as you please) was a French invention, the credit for bringing the steam engine to a high degree of automotive development goes to the British, among whom, incidentally, were numbered many of the world's most accomplished railroad locomotive designers. The great physicist and engineer James Watt began to apply himself circa 1765 to a scientific study of the

24

In what is said to be the world's first automobile accident,
the Cugnot crashes into a courtyard wall.

applications of the steam engine in the propulsion of road
vehicles. For some unknown reason, however, Watt never
concluded his studies of automotive problems, possibly because he became heavily engaged in large-scale production of
stationary industrial engines for textile plants, then mush-
rooming in England.

One of Watt's competitors, London merchant Francis
Moore, and Watt's workshop manager, William Murdock, took
the most positive steps of their time toward development of
the steam road carriage. Moore actually demonstrated a working steam carriage in front of the King of England, and he
obtained the royal approval. Murdock, possibly to keep his
accomplishments from the knowledge of his employer, Watt,
experimented at night and on out-of-the-way country roads.
During the early trials in 1784 there were enough accidents to
generate much unwanted publicity for Murdock, and he became locally prominent for a brief time. However, neither
Moore's nor Murdock's work produced anything more than
some interesting exhibits in the British Museum.

Auto Engines of Tomorrow

These men's successors, first among them being William Symington, brought Great Britain to the forefront of steam technology. Symington built a vehicle similar to Murdock's. The Symington machine received much attention from the public, some of whom could be best described as enthusiastic while others were simply fearful.

Murdock was also followed by Richard Trevithick, who built the first vehicle to operate on rails (a tiny train for hauling coal inside a mine) and constructed the first steam-powered threshing machine. Between 1796 and 1801 Trevithick built various tricycles for goods transport and in 1801 he built the first vehicle for passengers. The latter weighed eight tons and was used partly as a carnival display that persons afoot could race for a small payment. It had a maximum speed of 10 mph on the level but could maintain a steady 3.5 mph up a rather steep hill, and Trevithick rented it for towing carriages for joy rides much in the manner of today's hayrides. Later in life Trevithick was refused a government grant commonly used at that time to recognize special ability and, like his predecessor Cugnot, died in solitude in 1833.

In America, where water transportation was easier than road building, inventors interested themselves first in marine applications of the steam engine. The inventor of the nail-making machine, several steam engines, and the long-used vertical multi-tubular firebox boiler, Nathan Read, designed and built in 1790 a particularly well advanced steam engine for road application, but nothing came of it; and John Fitch in the late eighteenth century founded the first American company to build steam engines but never put any into production. A dedicated scientist and successful businessman of Pennsylvania, Oliver Evans, tried for many years to obtain a patent for a steam engine for road propulsion and for powering a flour mill. In 1804 he demonstrated publicly in Philadelphia an unusual amphibious vehicle that aroused much curiosity, but no records indicate any substantial success with it.

Meanwhile in England the steam carriage was becoming popular. More efficient engines had been developed for public road-vehicles. The outstanding pioneer in that endeavor was Julius Griffiths, inventor and builder of the first real bus to be put into regular public service for passenger transport. Griffiths' bus dates from 1822, a year during the period Robert Stephenson was operating the world's first steam railway. It is worth noting that early steam buses for the roads and locomotives for the rails were in parallel positions technically and, for a short while, economically and financially. In the field of commerce the two modes ended up in head-on collision, with the steam bus suffering total defeat in England. Powerful railroad investors were able to enlist the aid of the stagecoach interests in Parliament to cause throttling tolls to be placed on road-using steam buses and finally to put them off the road with the Red Flag Act of 1865.

Better roadbeds and other technical improvements in railroading by mid-century had greatly increased the competitive position of trains in England and elsewhere. Steam coaches (buses for the highway) were not legislated against or otherwise artificially hampered in France, yet they had begun a decline paralleling that in England. There were numerous attempts at revival of the great steam coaches in France during the time their use was restricted in England. Coach service between Paris and Versailles, begun in 1835 and said to be most comfortable because of the vehicles' specially cushioned wheels, was ended a couple of decades later. Amedée Bollée, a bell founder of Le Mans, built a 25-mph vehicle that was fitted with a gear changer and had independent front wheel suspension. Later his more advanced steam carriages "La Mancelle" and "Rapide" had some features characteristic of modern motor cars, such as a front-end engine (predecessors' engines usually were at rear or center positions) and rear driving wheels propelled by a driveshaft through a differential. The Bollée family is said to have made a small fortune selling these models, which were enthusiastically accepted in Germany and Austria as well as in France. An engineering officer in the Sardinian Army, Virgilio Bordino, copied an English coach in building the three steam buses known to have been produced in Italy.

Farsighted engineers fought with passionate zeal in attempts to secure the success of steam locomotion. Most of them thought big and produced coaches, carriages, and buses for types of transportation now euphemistically called "mass" and "rapid." They could not have known that the truly fabulous successes in the transportation field of the next hundred years lay in the field of *individual* transportation—a car (or two or three) for every family, and in many cases a car for every member of the family. But bear in mind that when the automobile as we think of it came upon the scene and quickly dominated the transportation picture it was a *small* machine compared to the huge buses that preceded it, a vehicle of only a few hundred pounds that could be propelled by two, five, or ten horsepower. A gasoline-fueled, internal-combustion engine was just right for that kind of automobile—it was cheap to build and therefore buy, used cheap and readily available fuel, and was cheap and simple to maintain. The internal-combustion engine fitted the needs of early automobiling. Electric didn't. Steam didn't.

But steam *almost* did. And many serious engineers think that our ancestors took a wrong course when they opted for internal combustion of gasoline rather than for the pickup, power, and pizzaz of steam or the instant start and the quiet and pollutionless propulsion of electrics.

We know what happened to the electric car—no one could come up with enough battery capacity. But what happened to steam? It was, after all, a medium of eminent success in stationary engines, river boats, ocean liners, locomotives, and yes, steam buses, which were technically successful but which had few good roads to travel on. Why couldn't steam make it in automobiles?

It's a question that has fascinated engineers from the

turn of the twentieth century, when gasoline, steam, and electricity seemed headed for a triple dead heat in the race for market supremacy. The superstitious have even felt that steam suffered from a jinx. Cugnot was thwarted by his government; Trevithick had too many accidents and bad luck in business; his English successors ran up against the insurmountable political odds promulgated by the railroad and stagecoach interests. Many auto historians believe that the first burst of steam was too far ahead of its time and that its revival came too late, at a time when electric cars were much easier to drive and maintain and gasoline cars were much cheaper.

The American identical-twin brothers, Francis E. and Freelan O. Stanley, built more steam cars than anyone else in history. In 1905, for example, they manufactured about 650 steamers; but the production totals of people like R. E. Olds and Henry Ford numbered in the thousands.

For several years Stanley Steamers were the world's fastest cars. As early as 1906 Fred Marriott, a Stanley mechanic, set five world speed records at Daytona Beach, Florida, once clocking 127.66 mph and becoming the first man to travel faster than two miles per minute. The next year he crashed on the beach at 197 mph and was seriously hurt. The Stanleys never raced their cars again, but their name was famous and their product remained popular enough to keep their plant busy. They continued to build 600 to 700 cars a year, until Francis died in 1917 and Freelan sold the company.

For all its fine workmanship, power, and speed, the Stanley had relatively simple machinery under its hood: a firetube, locomotive-type boiler set vertically and a 2-cylinder engine. The small furnace heated the water in the boiler to a very high pressure—the boiler could stand 1,300 pounds per square inch—and the resulting steam, after expanding in the cylinders and driving the pistons, was merely released to the atmosphere via an exhaust pipe. With no attempt to recover the spent steam and return the condensed water to the boiler, the Stanley Steamers—except some much later models—needed frequent replenishment of water. When the city governments of Chicago and Boston threatened to ban their steamers from the streets because of excessive steam discharge, the Stanley brothers relented and equipped a few cars with condensers.

The condenser, vital to all automotive steam engines of today, is a heat-exchanging device similar to the radiator of liquid-cooled gasoline engines. The spent steam, instead of being vented to the atmosphere to be lost forever, is directed from the exhausting cylinder to the condenser, where it is reduced in temperature enough to convert the vapor to water, whence it is pumped to the boiler to be reheated and started on another cycle.

The method involving the closed cycle is named for William John Macquorn Rankine, a Scottish scientist who died in 1872. Rankine won several prizes in physics while a student at the University of Edinburgh, then took up civil engineering and wrote a book on railway mechanics in 1842. He did research in molecular physics off and on for the rest of his life, was appointed to the chair of engineering at Glasgow University, and wrote several more books, among them *Manual of the Steam Engine and Other Prime Movers*.

The closed, or Rankine, cycle-method of steam-engine design has a number of advantages. It is quiet, without the typical exhaust "choo-choo" of the locomotive; it needs enough liquid for an initial system-fill only, with little or no after-adding; it is virtually nonpolluting; it can use a variety of fuels; and, in a modern design, it would be dependable, inexpensive to operate, and very powerful for its size.

The fabled locomotive that blasted its way across several continents and was a vital part of the development of the American West used an open system, rather than a Rankine cycle, for several reasons. The open system was simpler, required less plumbing, did not need what would have been a large and expensive condenser. Water was plentiful in most areas along most rights-of-way. There were no conservationists to object to the belching of smoke, cinders, and steam or to worry about the fire hazard from red-hot furnace products. Moreover, the steam exhaust performed a valuable job for which it would have been difficult to find a substitute: The exiting steam was directed up the stack in such a way as to pull after it a powerful draft that sucked fresh air into the furnace. The draft fanned the fire in proportion to the rate at which steam was being used, and it eliminated the need for a mechanically operated bellows arrangement that would have been complicated and expensive.

In large steam engines, such as those on locomotives, there are some formidable lubrication problems, and large quantities of lubricating oil are used. In practice, that oil becomes mixed with the spent steam that goes up the stack. Recovering spent steam would also have involved a filtering system to separate the oil from the water. Without it, the oil in the water being returned to the boiler would cause severe sludge problems.

The fact is that it simply was never considered necessary or even important to recover the water, particularly since it performed a service by chuffing up the stack and creating a useful furnace draft, and since the railroads already had adequate—in fact, enormous—water systems along their rights-of-way. At the end of the long era of railroad steam—when the diesels were beginning to take over in force—there were many large steam locomotives that carried as much as 20,000 gallons of water in their tenders yet required replenishment of water approximately every seventy-five miles.

In the Stanley and other steam cars with open systems, water stops were necessary every twenty-five miles or so, and this requirement was considered a major disadvantage. Rollin White, a son of the founder of White Sewing Machine Company, had a bit of engineering education and handled the maintenance of an 1899 Stanley that belonged to his family. Eventually he developed a semi-flash boiler (for quickly heating only a small part of the boiler's water) and other improvements that were successful enough to persuade the family to go into the steam-car manufacturing business.

The car they first built and marketed in 1901 had the appearance, prophetically, of a gasoline car, with a Rankine system including a condenser that looked like a car radiator. It thus did not have the water troubles of the Stanley. The White steamer was produced for about ten years and did well in the marketplace; but eventually the White family bowed to competition and began manufacturing an internal-combustion, gasoline-fueled engine for their car. Soon the White name moved from passenger cars to trucks and buses. White is one of the important pioneer names in American industry, living on as the brand name of one of the world's great modern sewing machines and one of the very top lines of heavy-duty trucks on today's highways.

In addition to the Stanley twins and the White family, several other steam car makers were in the business during steam's heyday, the teens up to the mid-twenties—MacDonald, Detroit, American, and Coats, to name a few.

Abner Doble, who made a total of forty-two steamers between 1913 and 1930, solved most of the problems of steam cars—burners that became sooted and needed frequent cleaning; boilers that became encrusted with salts from impure water; boiler tubes that blew out and had to be replaced; a water-level glass that frequently broke; the troublesome job of starting a fire under the boiler; the fifteen to thirty minutes it took to build up enough pressure to run the car; the water-level gauge that had to be examined frequently; the inconstant steam pressure involved in ascending and descending hills; the stop every twenty-five miles or so for additional water. Doble built a truly beautiful car that could go a thousand miles or more without replenishment of water. After he showed his car in 1917 he was swamped with more than $27 million worth of orders, but the World War I Emergency Board prevented his filling them due to material shortages.

Doble overbuilt his cars. He gave 100,000-mile warranties on the furnace-boiler system and three-year warranties on the remainder of the car. Some Dobles are reported to be running today, one with 800,000 miles on its odometer. But the price was $11,200 at the time Henry Ford was cranking out cars at less than half a grand. So without a Henry Ford to promote and produce, the steam car gradually gave up the small part of the market it once held.

In the early 1950s, R. P. McCulloch of chain-saw fame decided to build and market a steam car that would surpass even the Doble. He hired Doble himself (who had left the country in financial trouble and been moderately successful with steam trucks in England) as a consultant. But after about two years of work with "new steam" it occurred to McCulloch—a fantastically successful businessman—that it would be financial folly to attempt to compete with Detroit, so he pulled out of steam in 1954.

The most important aspect of steam had been completely overlooked until the late 1950s; and today it is that aspect—low to zero pollution—that gives steam its prominent place among engines of the future. Here's why: In an internal-combustion engine there is no way in our present technology to burn fuel correctly. Engines of today are far from efficient, but many users would be willing to waste. The problem is that the waste itself is poisonous to all of us. The intake, compression, combustion, and exhaust process of the world's most popular engines is not a good way to burn fuel, and improperly burned fuel generates exhaust products that we can no longer tolerate. So far, all we have been able to do is reburn or otherwise chemically change the exhaust products—at great expense—so they will not harm us as much as before.

In an external-combustion engine, however—and the steam engine is the most prominent example of an external-combustion engine—it is possible to impose precise regulation of the burning process, since it happens outside the engine and at near-atmospheric pressure. When we can burn a fuel purely for the purpose of providing heat, and we can call the shots with complete control, we can burn the fuel in such a way that it will be completely converted from a hydrocarbon-air combination to a number of harmless products such as carbon dioxide, water, carbon, and air.

This is not new information. In the late 1920s a defender of the steam engine pointed out in a magazine article that elimination of obnoxious gases would be welcomed as a wholesome boon. At that time no one paid any attention. It's hot copy now, and it has been a salable commodity for conservation writers for two decades.

More than conservationists are involved with steam now, however. Big business and government (federal and state) are in the act today, and from the rather prodigious spending of (mostly tax) money that's going on, something useful might emerge.

After McCulloch gave up his steam plans, not much happened for about a decade. There was some talk about electric cars, beginning in 1966, including the introduction of two Senate and two House bills that would have involved the government directly in research and development of electrics. But J. Herbert Holloman, who was at the time the assistant secretary of commerce for science and technology, testified at joint hearings of the Senate Commerce Committee and Public Works subcommittee on air and water pollution. He suggested that a panel be set up by the Commerce Technical Advisory Board to study the possibilities and indeed the potential of electrically powered vehicles. The resulting panel several times expanded its viewpoint and finally, in two volumes dated October and December 1967 it issued its report, *The Automobile and Air Pollution: A Program for Progress.*

The report pointed to the crisis in air pollution but said little to promote the electric car, which it called "limited." The panel did, however, say good things about steam. Among the sixteen panel members were Holloman, Richard S. Morse, the chairman, who was a government and industrial consultant associated with Massachusetts Institute of Technology and a former assistant secretary of the army, and David V. Ragone, who then was a professor of engineering at Carnegie-Mellon University and who is now dean of the College of Engineering of the University of Michigan at Ann Arbor. Morse and Ragone became so interested in steam as a result of their investigation that they and seven others formed Energy Systems, Inc. to manufacture steam engines. Even though the panel had concluded that there were serious difficulties with every other power source that had been proposed to replace the gasoline internal-combustion engine—but that the steam engine had possibilities as an alternative—nothing much came of the Energy Systems, Inc. effort in the steam field. Others have been involved in the attempted resurrection of steam; some of them are still at it; a few of them probably will succeed in bringing the public a successful departure from the gasoline internal-combustion engine. The chief question is *when*.

Edward Pritchard of Melbourne, Australia, believes that steam propulsion holds the answer to harmful emissions from automotive exhausts. He is the inventor and developer of a steam power system for automobiles that has been successfully installed and tested in a 1963 Ford Falcon. Pritchard, an honored member of the Institute of Engineers of Australia, also feels that steam power is competitive with other forms of automotive propulsion and in many ways superior to them. In July 1972, Pritchard was presented the Hartnett Award in recognition of his work on steam-driven automobiles by the Society of Automotive Engineers of Australasia in ceremonies at Australia's National Science Centre, at which time he voiced the conviction that the search for clean air makes the steam engine "the obvious choice for the future."

As early as 1968, Pritchard gave expert testimony before a joint committee hearing of the U.S. Senate on the perform-

Edward Pritchard, an Australian automotive engineer and the chairman of Pritchard Steam Power Pty. Ltd. of Melbourne, works under the hood of a conventional automobile in which his powerplant has been installed. (COURTESY OF PANCOASTAL, INC.)

ance of the Pritchard steam power system compared with the internal-combustion engine under stop-and-go traffic conditions:

> With the steam engine, a typical burner mixture is 10 percent lean—there is 10 percent more air in the mixture than is chemically required for combustion. With the car stopped as at a red traffic light, the engine is stopped. There is no idling and the burner, if of the on-off type [as in the Pritchard unit], is normally off. The whole system is dead when you stop, so there is no exhaust emission at all.

In 1970 Pancoastal, Inc. of Hartford, Connecticut, purchased a license option from Pritchard Steam Power Pty. Ltd., and early in 1972, after extensive tests in Melbourne and vicinity, exercised the option by acquiring American manufacturing, marketing, and licensing rights.

Within the space formerly occupied by a small internal-combustion engine, Pritchard installed a condensing steam power plant that utilizes a monotube steam generator and a full uniflow engine. The two cylinders of the reversible engine are arranged in a 90-degree V pattern. The engine, located behind the steam generator at the rear of the engine compartment, is fitted with "early" or "late" steam cutoff valve operation. It can be idled independently of the driveshaft.

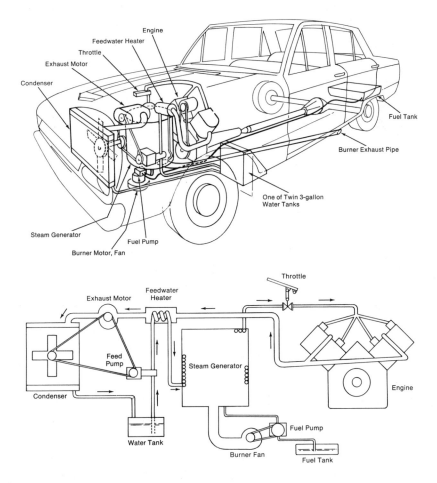

Installation diagram (top) and schematic diagram (bottom) of the Pritchard Steam Power System. (COURTESY OF PANCOASTAL, INC.)

There is no gearbox or automatic transmission, and although the engine (expander), separate from the burner-generator, weighs only 125 pounds, it is equal in power strokes to a conventional 8-cylinder gasoline engine.

The pistons are of the double-acting type (steam is fed to both ends of each piston) and, lacking side-thrust, require very little lubrication. Engine oil is injected into the steam supply at a point downstream from the throttle; it is filtered out of the condensate in the feedwater tanks. In the Pritchard system, roller bearings are used on the main journals and at the crank end of the connecting rods. These provide a minimum friction level and a design goal for engine life of 250,000 miles.

The steam generator consists of a monotube coil 13 inches in diameter covered by a 15-inch casing that is just 15.5 inches high. Its empty weight is 110 pounds. The generator is fired from the bottom by a burner aimed into a vertical combustion chamber. Exhaust is taken from the base of the generator. A 12-volt electric motor powers both the fuel pump and the burner's air blower, in the fashion of a common American oil furnace burner.

The fuel for the Pritchard system is kerosene or jet fuel, which is super-clean kerosene. The burner, when on, has only one firing rate. With the car moving slowly or stopped, the burner is usually not on for more than about five seconds, with fifteen-second or longer intervals of off-time in between. The burner will tend to stay on during acceleration or high cruising speed, and it normally goes off when the throttle is closed (foot off the accelerator), as during deceleration for a stop; it then remains off while the car is stopped, with zero exhaust emissions. The working fluid is water; for cold-weather operation the feedwater will have to be compounded with a very stable antifreeze. The company was working on that, plus a number of other refinements for their system, during 1974 and had not announced a beginning date for production of road-ready cars.

Others in the forefront of steam-car research are—no order of importance intended—Lear Motors Corp. of Leareno, Nevada, an installation on a former air base near Reno; Scientific Energy Systems of Watertown, Massachusetts; Thermo Electron Corp., Waltham, Massachusetts; and Aerojet Liquid Rocket Co., Sacramento, California.

Smaller cars and necessarily compact powerplants are not just a trend, in the opinion of Jack H. Vernon, president of Scientific Energy Systems. Accordingly, development by his company of a Rankine-cycle steam engine for our future automobiles is an effort in the right direction, he feels. SES has fitted its latest equipment—boiler, expander (engine), condenser, and all accoutrements—*plus* air conditioning, power steering, and power brakes, into the engine compartment of a Plymouth Fury, leaving an air space of nearly twelve inches both fore and aft. The SES setup puts out about 150 hp, not much for a steam engine but possibly enough, as Vernon claims, to push a 4,000-pound car. The warm-up capability is in the neighborhood of thirty seconds or less. Although the SES engine can use any kind of cheap fuel, low-octane gasoline down to a kerosene-grade oil, even the promoter claims only about eleven miles per gallon, which is average-to-good for a late-model conventional car of comparable size. Vernon emphasizes, however, that further refinements will both reduce weight and size of the powerplant and increase fuel economy.

It surprises no one that the different modern investigators into the possibilities of steam have begun to arrive at different conclusions. The SES plan is to concentrate on pure (that is, demineralized) water as the working fluid, and to use that medium in a package that is similar to a 4-cylinder diesel engine.

Bearing in mind that water, the earliest and to this day the most popular working fluid, is inorganic (containing no carbon), note that Thermo Electron is using organic expander fluid (a type of fluorocarbon) in a piston engine, Lear is using water in a turbine (after disappointments with the piston), and Aerojet is attempting, with considerable success, to combine an organic fluid with a turbine. This means that SES is the only major company using water as an expanding fluid in a piston engine, in the tradition of Cugnot, the Stanleys, and Doble. The danger of damage to the engine and inconvenience to the driver because of freezing is handled, not with the addition of alcohol to the working fluid, but with a method of draining the water into an insulated sump and keeping it warm when necessary by means of a pilot flame.

The efforts of SES and others have been spurred by the 1975–76 exhaust emissions standards promulgated by the Environmental Protection Agency. The ability to meet the standards and still furnish the power we require in our cars has been the object of most engine research and development during the past decade, 1965–1975.

During 1972, SES successfully negotiated $2.1 million for additional work under its contract with the Environmental Protection Agency. This allowed continuation of development of the SES "Rankine cycle automotive propulsion system," as the company prefers to call it. For several years SES has been developing and testing its prototype 150-hp engine system in its laboratory. The company has successfully installed a mock-up in a redesigned, full-size 1974 Dodge. Road testing of a completed demonstration model is planned for 1975.

The expander (steam engine) is a 4-cylinder, reciprocating (piston) type with the pistons, bearings, and cam surfaces lubricated by crankcase oil. Auxiliary and accessory equipment (alternator, etc.) are belt-driven by the expander, which, as in conventional automobile powerplants, idles when the vehicle is stopped. Due to full balancing, the expander is designed to run at relatively low vibration and noise levels.

SES has proprietary rights to a highly compact burner-steam generator that provides quick start-up and rapid operating response. The manufacturer claims that the system

Mock-up of a modern 150-horsepower steam engine installed
in a redesigned, full-size 1974 Dodge engine compartment.
(COURTESY OF SCIENTIFIC ENERGY SYSTEMS CORP.)

operates with very low levels of exhaust emissions compared
to conventional gasoline and diesel powerplants. Unburned
hydrocarbon emissions have been as low as 8 percent of the
maximum allowed under the 1975 federal standards, and
carbon monoxide as low as 10 percent. Most significant, perhaps, is the reduction of oxides of nitrogen (NOx) to about
50 percent of the amount allowed in the 1976 standards. The
NOx emissions are the most difficult to manage in internal-
combustion engines.

The SES system uses six quarts of water as the working
fluid. Water has excellent thermodynamic properties so that
it carries and exchanges heat easily. Exhaust steam from the
engine is sent to an air-cooled condenser similar to an auto
radiator. Here a dual fan system exchanges heat from the
spent steam, converting it to water, and a special pump re-
cycles the water back to the boiler. Esso Research and Engineering Laboratories, a subcontractor to SES, has developed

A torn-down view of a V-4 Rankine-cycle engine block and piston-connecting rod assembly. (COURTESY OF THERMO ELECTRON CORP.)

a suitable lubricant for steam operation, and Bendix Research Laboratories, another subcontractor, has satisfactorily completed the first phases of development work on systems controls. As part of its subcontract to provide vehicle engineering assistance, Chrysler Corporation built a full-scale mock-up of the SES system in an actual automobile engine compartment. That mock-up was displayed at the December 1972 EPA contractor coordination meeting at Ann Arbor, Michigan.

Thermo Electron organic Rankine engines have done well in the area of low exhaust pollution and improved energy utilization, but it remains to be seen whether the organic working fluid has advantages that surpass its competitors for future good position in the marketplace. The company gained increased support from the federal government in 1973 with the award of an additional $1.1 million by the EPA. Through dynamometer testing, TE confirmed the extremely low emissions that it had theoretically projected with its 150-hp organic Rankine engine. In continuing work with Ford Motor Company and the EPA, TE is striving toward further improvements in both performances and fuel economy. During 1973 the company delivered prototype organic Rankine engines to an unidentified Japanese manufacturer of small

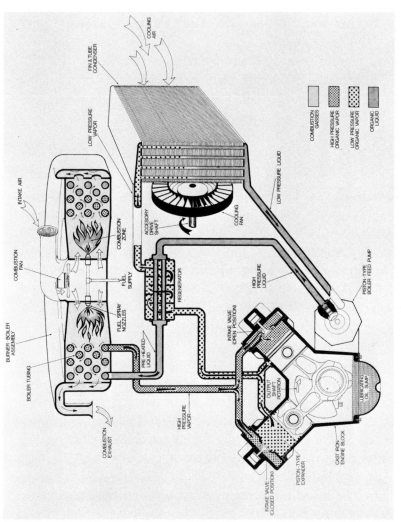

Schematic diagram of the Rankine-cycle engine developed by Thermo Electron Corporation. (COURTESY OF THERMO ELECTRON CORP.)

43

vehicles, and both reciprocating and turbine versions of TE's small Rankine engine were evaluated in the testing program in Japan.

The major system components of the completely sealed Rankine-cycle engine on which TE is concentrating its efforts are a burner-boiler, expander, feed pump, condenser, and regenerator. The regenerator is a special heat exchanger that provides close association between the spent vapor just exhausted from the expander and the liquid on its way to the boiler. The regenerator cools the former as it is on its way to the condenser, and it warms the latter as it is on its way to the boiler.

In operation of the complete system, organic fluid is heated in the boiler and sent in the form of vapor to the expander, where it does the work of powering the engine to drive the vehicle and run all auxiliaries and accessories. Superheated vapor exhausted from the expander goes to the regenerator where part of the energy remaining in the vapor is transferred to the feed liquid that is on its way to the boiler. The vapor is then sent to the condenser, where fan-driven air removes the remaining superheat, condensing the vapor to a liquid form. The condensed liquid's pressure is then raised— from condenser pressure to boiler-inlet pressure—by the system's feed pump. The liquid then passes through the liquid side of the regenerator, picking up some heat, and on into the vapor generator (boiler). At this point, energy is transferred to the working fluid as combustion in the burner raises its temperature and changes it to a vapor capable of running the engine. A new cycle has thus begun.

The working fluid employed by Thermo Electron is Fluorinol-85, known to chemists as a mixture of 85 mole percent trifluoroethanol and 15 mole percent water. The company considers it to have excellent characteristics as the working fluid for low-cost Rankine-cycle engines oriented to commercial applications. Its thermodynamic properties provide an acceptable cycle efficiency at moderate boiler output temperature levels of 500 to 625 degrees Fahrenheit. At such relatively low temperatures, low-cost lubricants are available that are thermally stable and compatible with the working fluid at the peak cycle temperature. Thus a lubricant can be mixed with the working fluid and sealed with it inside the system, providing good lubrication for the internal moving parts, an essential for long operating life. Since the working fluid is noncorrosive and kept at a moderate temperature, low-cost construction materials, such as ordinary carbon steel boiler tubing and regular cast iron for the engine block, can be used to keep costs involved in the entire system to a minimum. Another plus is the –82 degrees F. freezing point of the organic working fluid.

With only minor modification of the vehicle, an early complete system was packaged into a 1972 Ford Galaxie, accomplished with the assistance of Ford Motor Company. The theoretical performance calculations for that installation called for 146 hp gross at the crankshaft to be produced at a vehicle speed of 107 mph and an expander speed of 1,800 rpm (less than one-half the usual shaft speed of a large V-8 gasoline engine at that vehicle speed).

"If there were no energy crisis," said William Lear in February of 1974, "we'd have the world by the tail on a downhill slide." Bill Lear is the only man in history who had the resources and the inclination to spend a vast fortune on the development of steam. And just as he seemed to be getting close to his goal—a steam turbine (he had already tried and abandoned a reciprocating expander) that uses water (he modestly calls it "Learium") as the working fluid—Lear found himself with a comparatively inefficient engine and confronted with a fuel shortage.

"But even so," added Lear "acceptance of the steam turbine *will come about* because of the higher cost and shortage of gasoline." Lear has invested more than $17 million of his own money in what is probably the largest personal automotive project ever undertaken. In 1974, Lear calculated that something in the neighborhood of $10 million *more* would be required to complete his project.

"We get about 60 percent efficiency from the steam turbine now, and we need about 75 percent efficiency. That doesn't seem like a lot, but it is. It's easy to make a very large steam turbine efficient but not so with a small one. Still, it can and *will* be done." In early 1974, he figured he would have the job done in about two more years.

The legendary Lear, who made several fortunes in a long and active career as a patenting designer of the Majestic radio, a cofounder of Motorola, and the developer of direction finders and other electronics equipment for aircraft in World War II, was awarded the Collier Trophy by President Truman in 1949 for his work on automatic pilots. Lear also put together Lear Aviation. That became the $200 million-a-year Lear Siegler, which he sold after arguments with his managers. Then he formed Lear Jet, the firm that produced, before 1970, more than two hundred of "the most economic, fastest, highest flying, lowest cost business jet ever developed," to use Lear's own enthusiastic but accurate description. Lear sold Lear Jet in 1966 to Gates Rubber Company of Denver for an excellent figure, went into a physical decline and a mental despondence, snapped out of both and surged into steam.

Today, in an old air-base property some ten miles north of Reno, Nevada (Lear calls the place Leareno), the man is devotedly spending money in pursuit of an engine that will not pollute the air and that hopefully will someday be efficient enough to compete with the conventional gasoline engine. After moving in and renovating old buildings in 1968, he started building additional facilities. In 1969 he added 600,-000 square feet of plant space, and he has never stopped expanding.

Lear believes that the first applications of his steam turbine (basically, a fan that turns inside a chamber when blasted by steam from a nozzle), when perfected for road use, will be in heavy-duty commercial vehicles—trucks, buses, tractors, perhaps even railroad locomotives—the same route followed by diesel engines before they were used in automobiles. Lear Motors Corporation has developed a prototype steam turbine-powered transport bus that began its testing phase in San Francisco in 1973. It was used in regular operations by the San Francisco transit system, and results were favorable, according to Lear's reports.

"Everyone knows [the steam-turbine engine] doesn't pollute, but we proved in San Francisco that it can do the same job a diesel can," said Lear in 1974. "It worked very well, but we are concerned with the efficiency." Lear's steam bus also made a demonstration run in Los Angeles, but no plans were shaped up to put a unit into regular commercial operation.

The Lear steam turbine uses water heated to a vapor to blast through a nozzle against a turbine wheel (a type of fan) to turn it and the shaft to which it is attached at a very high speed—approximately 65,000 rpm—which via gear reduction is greatly lowered to be suited to the driveshaft speed of the vehicle.

The water can be heated by any fuel, but for convenience and economy at the present time, diesel oil is perferred. Since that fuel (or almost any other) burns more completely, efficiently, and cleanly in the type of furnace (at near-

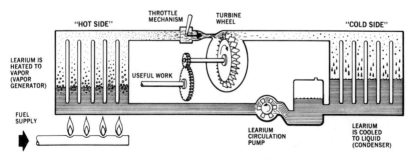

Greatly simplified schematic drawing of the Lear Vapor Turbine System. The "working fluid," here called "Learium," is chemically identical to water. (COURTESY OF LEAR MOTORS CORP.)

atmospheric pressure) employed by the Lear system, pollution levels are much lower than in an internal-combustion engine. Unfortunately, fuel consumption is high compared to the automobile gasoline engine and especially when compared with the consumption of the diesel engine, the most economical of all popular powerplants.

The project of working the bugs out of the steam turbine takes up to 90 percent of Lear Motors' time and resources. The company keeps its financial head out of water by producing and selling Lear's many different patented products, said to number about 350. Lear is uncertain how his company will market the steam turbine once it is perfected. Possibly, Lear Motors will issue stock to raise enough money to manufacture a complete steam car.

"If market conditions were such as to accept a large public issue, we would go ahead. But that condition simply doesn't exist at the present time," Lear said in 1974. "We've spent some $17 million developing the steam turbine, and we just don't have enough to get into the manufacturing of cars. It would take about $100 million to set the whole thing up."

There's the possibility, of course, that Detroit will be forced to turn away from the internal-combustion engine and adopt the steam turbine or some other alternative. But Lear feels that unless Detroit is forced by public outcry into abandoning the gasoline-piston engine, it will never accept the steam engine. "They wouldn't believe in the steam turbine if it ran over them," Lear said, referring to the Detroit establishment. "Dealing with the auto industry has been the most disappointing experience of my life. They just don't want to go to steam. They've got millions of dollars tied up in tooling for the piston engine, and adopting the steam turbine would upset all that."

In an early 1974 newspaper interview, Lear also revealed some of his inner thoughts about the federal government's involvement in automotive antipollution matters.

Expressing little faith that Washington might move to speed development and acceptance of the steam turbine, Lear concluded: "Look at the EPA's phone book sometime—it's enormous. They've spent their money building a huge bureaucracy instead of attacking the pollution problem. I have no confidence in the EPA's competence."

Lower smog-producing emissions are a forceful characteristic of two experimental automobiles built as prototypes in a $2.8 million state-financed program to develop low-fuel-consumption vehicles in California. So far, the vehicles have not solved any fuel-shortage problems, but they are low in pollution production. The technical manager of the program, Roy Renner, reported in 1973 in an interim progress study that it may eventually cost private industry hundreds of millions of dollars_to improve gas mileage in mass production of cars similar to the prototypes then under consideration.

By that time, slightly more than $450,000 had been invested in the project with an addition of more than $1 million still to go into the effort. The work was being done under a $1.46 million contract negotiated by the California legislature with Aerojet Liquid Rocket Company of Sacramento and a $1.36 million contract with Steam Power Systems, Inc., San Diego. The funds came from a legislature contingency kitty not subject to control by any source other than the legislature itself.

Two subcompact experimental steam-power cars built under the contracts were designed to operate in city traffic at a fuel consumption rate of 10 to 12 mpg and up to 19 mpg at sustained cruising speeds of 50 mph. "This compares rather poorly," said Renner in 1973, "with the 16 to 18 mpg that can be obtained from a standard Chevrolet Vega or with the Volkswagen's 18 to 20 mpg in city driving."

Renner continued: "In terms of smog emissions, the cars with Rankine-cycle steam engines will release into the air slightly less than one-half the emissions permitted under U.S.

standards set for 1976." Those were the standards that Detroit auto makers argued they would not be able to comply with by the target date.

However, Renner added: "Once the emission problems have all been solved, private enterprise surely should be able to handle the second-generation research necessary in solving the problem of relatively poor fuel mileage."

When the project was started in 1972, economical gasoline consumption was not considered a major priority compared with the need to devise a method of achieving low engine emissions without loss of driveability.

"We are now working, hopefully," Renner said, "on increasing mileage by means of transmission refinements and other improvements that can be built into the experimental car in time for the scheduled 1974 delivery date."

Since these things never seem to happen on time, is it unfair to wonder whether the public relations departments of the companies so deeply involved with steam are not themselves blowing off a bit of steam?

3

The Motor That Could; the Battery That Couldn't

A DETROIT group of automotive experts and executives, meeting in Cleveland in 1973, thought they were hearing things. Before them stood Semon E. Knudsen, board chairman of White Motor Co., one of the world's top producers of transport trucks ($1 billion worth per year). Knudsen had spent many years as a General Motors executive, had been manager of Pontiac and manager of Chevrolet, and had been a strong contender for the GM presidency. Later, before going with White, he had been president of Ford.

Knudsen told the astonished group: "We have been hung up too long by our preoccupation with the reciprocating engine." He was referring to the powerplant of the "gas buggy" that R. E. Olds began mass-producing early in this century and that is currently reproduced at the rate of about ten million per year—the conventional gasoline-fueled piston engine.

In that possibly historic speech, Knudsen came out strongly for the electric vehicle, the EV, as a hopeful replacement for the car of today. Knudsen is as well aware of the problems inherent in EV production and marketing as anyone, and more so than most, but he seems willing to face up to the fact that pollution problems coupled with the current fuel-consumption emergency call for a drastic reevalution of alternatives to combustion-powered vehicles.

51

Auto Engines of Tomorrow

Technically speaking, even the EV is a combustion-powered vehicle, in that the electrical power used to recharge its batteries is often produced at a power plant that burns coal, oil, or both. There is also a sizable percentage of our power, however, that is produced at hydroelectric installations, and no combustion is involved in such production. Serious engineers of today believe that if the EV had caught on and progressed early in its history as the gasoline engine did, we would now have a much larger hydroelectric capacity, which is pollutionless.

Knudsen spoke many months before the great energy crisis of 1973 hit the country, and there was prophetic thought behind his words. He indicated that he was strongly opposed to a continued reliance on fossil fuel (coal and oil) as a power source. He suggested that Congress might eventually have to pass a law requiring auto manufacturing companies to develop new kinds of motive power. As an example, said Knudsen, Congress might have to tell car makers that by, say, 1985 they must invent and put into production a small and efficient storage battery (similar to but much larger than the car starter battery of today) that could power an automobile for 500 miles without a recharge and that would last indefinitely. That would be a tall order, to be sure, since nothing approaching even 25 percent of such capability exists in the most up-to-date technology. But Knudsen feels that an advancement of those dimensions is not only possible but mandatory to the preservation of the ecology.

Storing electricity (that is, electrical power) "may be the most important single challenge of our century," Knudsen told the Detroit people. Solution of the energy crisis, he feels, is a responsibility of the motor-vehicle industry. "What we need above all," he said, "is to set aside substantial amounts of money for research aimed at finding new kinds of motive power." It is to his credit that he does not feel that the government should come forward with the development money. He recognizes that the energy crisis is of far greater importance to our future than the auto industry's problems with exhaust emissions. And he feels that every effort should be made, immediately, to find a replacement for petroleum as a fuel. To justify such a move, Knudsen referred to the rapid increases in consumption of petroleum and natural gas concomitant with a tightening of supply that could produce a "colossal" problem. He added:

> It is up to us [automotive industry people] to anticipate national problems that involve transportation and not wait for others to identify the problems and prescribe the solutions. We certainly owe it to ourselves to avoid the errors of the past, when we allowed a vacuum to develop with regard to both automotive safety and air pollution. We can no longer say we are too busy with our day-to-day problems to bother with national issues. If we take that course, we [industry members] will be attacked once again—and this time we could be destroyed.

These were the words of a rich, prominent, powerful automotive industrialist who said he felt that if car makers ignore the energy crisis and the need for vehicle powerplants that don't require petroleum fuel, the result might well be the end of the free enterprise auto industry as it exists today.

As mentioned in Chapter One, the first automobile track race in America was won by an electric, a Stanhope made by Riker. Seven vehicles were entered, two electrics and five Duryea "gas buggies," but two of the Duryeas were disqualified. The Riker won all five heats and averaged 26.8 mph. Second prize went to the other electric, a Morris and Salom Electrobat. An authentic historical account indicates that this 1896 race was so dull some of the spectators started yelling, "Get a horse," probably the origin of the early-day automotive imperative.

Of even greater interest to electric-car buffs is the fact that the electric was the first car to go a mile a minute, speed never having been one of the electric's more compelling features. Yet some eight months before the turn of the century, in a suburb of Paris, Camille Jenatzy drove his *La Jamais Contente* to a world land speed record by clocking an amazing 65.8 mph. The car was designed with every known deference to aerodynamic principles—it came to a sharp point in front and looked like an aerial bomb lying on two two-wheeled axles —and could stand as a good lesson to the battery-car builders of today who are right now fighting the physical fact of wind resistance varying as to the square of ground speed. The nature of the electric motor and the lead-acid battery was such as to dictate that an electric car could go fast for a very short time, or go much slower for a much longer time. That was the problem then. That is the problem today. And although the solution is not in sight, many thinking people feel that the solution should be unstintingly pursued.

In 1899, more steam cars (1,681) were produced in the United States than electrics (1,575) or gasoline cars (936). Ahead of gasoline in number of cars produced, and running neck and neck against each other, steam and electric power seemed to be vying for automotive dominance. Just after the turn of the century, the electric was apparently firmly established as the most popular mode of automobile transportation. An electric had first appeared in Chicago in 1892, and at that time it created quite a furor. A newspaper story had this coverage: "In a number of instances so great has been the curiosity of those on the streets that the owner, when passing through the business section of the city, has had to appeal to the police to aid him in clearing the way for his carriage." Electric cabs were common in New York and other large cities by 1900; and even earlier, electric delivery wagons were put into use by New York and Boston stores. In 1905 at the Fifth National Automobile Show, there were displayed 177 different gasoline models, 4 steam cars, and 31 electrics, 9 of which were trucks.

By 1908 there was even a "hybrid" electric car, the Columbia Mark XLVI, which had a gasoline engine that drove an electric generator. The generator powered separate electric motors that were connected to the wheels. The method—similar to that used on many diesel-electric locomotives of today —was designed to provide a smooth flow of power without need for a conventional gear-shifting transmission.

By 1912 electric-car manufacturers were organized to hold separate automobile shows, and in 1914 the Society of Automobile Engineers (predecessor of today's SAE) organized several research committees to study the advantages and disadvantages of electric vehicles. In 1924, for the first time, not one electric car was displayed at the Twenty-fourth National Automobile Show.

The electric car had one outstanding quality that never was and possibly never will be claimed by gasoline or steam —utter simplicity. It had another quality that it could certainly lay claim to in comparison with its contemporaries: dependability. Moreover, the electrics quickly became the luxury autos of their day. The Pope of 1905 was advertised as "A Princely Gift for Your Wife," and the Baker Electrics of 1909 were described by their maker as "The Most Luxurious Ever Built" and "The Aristocrats of Motordom." All the electrics, whether the sales leaders like Pope and Baker, or the other famous marques like Argo, Columbus, or Rausch and Lang, had special appeal for ladies. There were never any starting problems regardless of the weather or climate, there was no complicated shifting or clutch action, and the car was so easy to drive that accidents with them were rare. Because they traveled slowly and almost always on paved city streets, the electrics could be fitted with solid rubber tires, which eliminated problems with punctures or blowouts. The electrics reached their highest point of development, popularity, and elegance in the period just before World War I.

With all their advantages of simplicity of construction and convenience and ease of operation, the electrics had serious disadvantages, most of which centered around their power supply. Their speed was low—rarely over 25 mph. Their range on a fully charged set of seven or eight very large batteries was only about twenty-five miles when electrics first came on the market. Remarkably, that range was tripled in less than ten years of production. But then, peculiarly, the range was not extended farther in the years from 1910 to 1918, and even today range shows little improvement although top speed has been increased in modern vehicles. Meanwhile, the early horseless carriages that burned gasoline could go more than a hundred miles on a tankful and then had to stop only a few minutes for a "fill up," while the early steam cars were gradually improved so that frequent stops for water were eliminated.

The second major problem with the electric—one that is nowhere near a solution today—was the frustrating amount of time required to recharge a set of "low" batteries. This was usually in the neighborhood of twelve hours, a so-called "overnight" charge that was all right for the housewife who used the car only for daily errands and could plug in her car at suppertime and do without it until the next morning. But the situation was hopeless for any kind of interurban travel—the car could be operated for two or three hours and then it had to be recharged for about half a day. If the electric car had become popular enough, if something on the order of 2 or 3 percent of today's cars were electrics, a large battery rental and exchange service would probably have developed, so that a traveler could stop when his car's batteries were low and exchange them for a fully charged set for a nominal fee, instead of having to wait while charging was done.

In the old days, owners of electric cars needed elaborate and expensive charging equipment, which was usually installed in the garage, to maintain the power supply in their vehicle. Quite different from the compact and relatively cheap transforming and rectifying devices available today, the old "charging panels" had ammeters to indicate the rate of charge, voltmeters to show the state of charge, and heavy-duty switches and large cables for transmitting the large charges required by electric cars.

During World War II there were still scores of the old

electric cars around, out of use but stored by loving owners in garages and barns across the nation. Many of the owners were delighted to discover that the cars could be overhauled and put back into service during the gasoline-rationing period. In England during that war, a manufacturer brought out small electric buses that could carry twelve passengers and their luggage. And because of the high price of gasoline in England, the good streets and roads, and the short trips involved compared to the longer distances in the United States, electric vehicles of several types have come into widespread use, especially in the delivery services, trash removal, and warehouse work.

In the United States, electric vehicles have found themselves confined to only a few areas up to the mid-1970s. In applications where slow speed is actually desirable or even minimum pollution is intolerable, the EV is unsurpassed. Thus we see an overwhelming popularity of the EV as a golf cart, as a messenger car in large offices, and as a forklift in warehouses. On the golf course, safety considerations mandate slow speed, and here the EV is much admired also for its quiet operation—it can be driven right up to the green without disturbing other players. It is the preferred type of vehicle on golf courses around the world and in fact is shunted aside in favor of the gasoline putt-putt type only where the terrain is difficult—EVs are poor hill climbers.

Except for the above-noted specific uses, the EV was considered totally obsolete in the period from 1930 to 1960. Unquestionably it would have stayed obsolete if there had been no pollution crisis and no fuel crisis. But the two crises—pollution in the 1960s and fuel in the 1970s—compounded to re-create an interesting market for EVs.

In the middle 1970s as this book is being written, there exists the technology to produce a small electric car that will travel about 60 mph for about two hours, or at slower speeds for longer periods of time not to exceed eight hours, after which its batteries must be recharged for six to eight hours.

Based on that technology, there are more than 40,000 EVs in the United Kingdom used for delivery of dairy and bakery products, and dozens of manufacturing companies have recently commenced EV production in the United States. To mention examples of a few types of EV developments, Battronic Truck Corp. of Boyertown, Pennsylvania, makes 25-passenger battery-electric buses and several types of delivery vans; an attractive little sports-type car, the Electra Spider, is made by Die Mesh Corp., Pelham, New York with a top speed of 55 mph and a range of 35 miles.

An electric car designed by a New Zealand scientist for commuting would run from the owner's home at speeds up to 30 mph on its own batteries until it arrived at a main road, where it would pick up an underground aluminum grid that would double the car's speed and recharge its batteries as it was propelled along, until time to leave the main road for the driver's destination, again on the self-carried batteries. In Germany, it is estimated that two million EVs will be on the road by 1985, with a network of battery-charging and -changing stations probably hooked to their own special power plants for recharging-power needs.

In Japan, the Toyota Townspider was demonstrated at a 1973 auto show with an innovative system for its operation: A subscriber would use a magnetic credit card to open the door of the Townspider, which would be parked in a special system lot. Then he would simply drive off and leave the car in another lot near his destination. The car would then be recharged and made ready for its next customer. A huge 80-passenger bus in Nagoya, Japan, is good for 110 miles at 30 mph before recharging.

In early 1974 an electric car of extremely light weight was introduced in Turin, Italy. Researched and built by Gianni Roglianni and designed by Giovanni Michelotti, the car has an aluminum body and a chassis that is mostly aluminum with occasional steel reinforcement. Reduction of gross vehicle weight by careful design in this case not only cut down on power requirements but also allowed the vehicle to operate on a smaller number of batteries, thus again cutting down on weight. This reverses the snowballing effect common in the electric vehicle field, in which normally more-batteries-for-more-power cause an increase in vehicle weight, which calls for more battery power, and so on.

The electric motor of the little Italian car is rated at 4 hp at an input of 150 amperes. Overall length is 92.5 inches, width is 53 inches, and height is 54 inches. The car without the batteries is only 770 pounds, making it lighter than many motorcycles; but the batteries weigh almost half as much as the car and bring total vehicle weight up to 1,122 pounds without passengers. Maximum speed is a respectable 37 mph, but range is a disappointing 30 miles.

Zagato of Italy introduced two small electric cars in early 1974, the Zele 1000 and its more powerful running mate, the Zele 2000. The Zele 1000 is mostly constructed of synthetic materials with a centrally located electric motor powered by eight 12-volt batteries. It develops 4.8 hp at 2,300 rpm, which produces a top speed of 25 mph and a range of nearly 45 miles. It takes about eight hours to recharge the battery. The Zele 2000 has a maximum speed of 35 mph but its range is only 30 miles.

One of the most interesting of the estimated 80 to 85 EV prototypes under test around the world in 1974 was a car built by the Electric Vehicle Group at Flinders University in South Australia. Whereas much time and money continue to be spent on the search for a new high-energy battery of light weight, the Flinders group is going ahead on the premise that no such innovation is on the horizon and that existing low-cost lead-acid batteries can be used to power their radically different type of electric car.

The Flinders design employs a motor that runs at a constant speed regardless of the car's speed. Vehicle speed-changes are handled by a variable-ratio type of transmission that actually is an oil pump. It is connected through pipes to hydraulic "motors"—to use the word in a sense different from "electric motors"—that are supplied with a pressurized flow of oil that makes them turn like turbines. There is one such "hydrostatic motor" at each rear wheel, and each motor is connected to its wheel through appropriate gear reduction.

A control system, operated by a foot accelerator, contains an electronic sensing unit, and the accelerator position at any given time causes the sensing unit to determine the required transmission ratio for moving the vehicle at the speed desired by the driver.

The electric motor that converts electrical power from the battery to mechanical (oil) pressure is a special type that is not familiar even to most electrical engineers. Capable of handling an input of 10 kilowatts (about 13.5 hp, neglecting heat and bearing-friction losses), the motor is equipped with what is called a "printed-circuit" armature. Its field is produced by permanent magnets that are not wound as are the field-coil electromagnets in a conventional motor. The armature itself, instead of being a group of wire-wound electromagnets bound around a shaft, is made from flat copper strips that are interleaved in four layers. Current flowing through the copper strips of this unusual type of armature sets up a magnetic pressure, called magnetic flux, that bucks, or pushes against, the standing magnetic field of the field magnets. The magnetic pressure spins the armature.

The air gap between the rotating armature and the stationary field magnets is so small that the armature can set up its flux without benefit of any magnetic material, such as iron, the usual component of an electromagnet. With no iron, the armature has very low inductance (unwanted, impeding magnetic forces). The flat-strip copper conductors are machined so that the motor's brushes (contactors made of graphite) can bear directly upon them, so that no commutator is needed, and the method virtually eliminates sparking at the brushes, an important cause of brush wear.

The absence of both iron and windings gives the armature

Chassis configuration (top), engine compartment (bottom left), and hydrostatic motor-wheel assembly (bottom right) of the Flinders Group electric vehicle. (COURTESY OF *AUTOMOTIVE NEWS*)

two salutary characteristics: low weight and low production cost. Another advantage is that instead of electronic speed controls that can be expensive and troublesome, only a simple on-off switch is needed between the battery and the motor. A drawback is the armature's low thermal inertia (it heats up quickly) compared to the heat-sink (thermal absorption) capability of the iron in the conventional electromagnet type of armature. This means that the Flinders motor must not be overloaded; but with the constant-speed, hydraulic-link operation for which this type of motor was designed, even a temporary overload is not likely.

The Flinders vehicle, designed for the commuter market, weighs 1,210 pounds. The lightweight body, vacuum-formed from ABS (acrylonitrile-butadiene-styrene) plastic, has seating for two adults and two children. The car has a range of more than 50 miles and a top speed of 40 mph. The Flinders group is pleased about the fact that their motor can be powered by a battery having very thin grids, allowing a substantial saving of weight and space, and providing a greater area of active material to be presented to the electrolyte. Tests show, also, that a greatly extended battery life is possible, due to allowable adjustments in the specific gravity of the acid-water mixture. The present Flinders battery is a 150-volt size that weighs 594 pounds, almost one-half the total weight of the vehicle.

In the United States, environmental and (more recently) energy concerns have been whipping up interest in EVs for several years. Remember that the first of the EV pioneers in this country did their work long before the last century was out: Frederick M. Kimball of Boston, in 1888, and William Morrison of Des Moines, in 1891. By 1912 more than 6,000 electric passenger cars and more than 4,000 electric trucks were being manufactured annually.

As of 1975 that kind of production had not yet been re-established, but it appeared that the nation was getting close to the old 1912 production goal. Sebring-Vanguard, Inc. of Florida had already begun production of its Citicar, displayed early in 1974 at the Electric Vehicle Symposium in Washington, D.C. with a retail price tag of $2,269, including a vinyl roof and certain required options for on-road use, such as custom side windows and speedometer-odometer. The Citicar has a two-passenger body with 12 cubic feet of luggage capacity behind the passenger seats where up to 200 pounds of cargo can be carried without adversely affecting performance. The vehicle is capable of 28 mph with a range of up to 50 miles on an overnight charge.

At the outset of production, Sebring-Vanguard was granted a one-year exemption by the National Highway Traffic Safety Administration (NHTSA) from four passenger-car safety standards: production was allowed without a defrosting system (although the car does have a fan and scoop system for defogging); without self-canceling turn signals, without conforming door latches and hinges, and without upper torso restraints (harnesses).

The Citicar is not equipped with a heater, and that brings up a formidable problem: to heat with electricity would be a severe drain on the batteries that would limit the mileage range to a small fraction of the normal, already-too-short range. Ironically, heaters fueled by gasoline have been proposed for electric cars; such devices had middling success on conventional cars in the 1930s. Their advantage was immediate heat; high cost was their chief minus factor. When thermostats were introduced to restrict flow of coolant from the engine block to the radiator until the engine was warmed up, conventional "hot water" heaters began to put out heat in two to three minutes, and gasoline heaters faded from the automotive scene. But we may see them back if a practical alternative in electric cars is not developed soon.

In its first year of manufacturing the Citicar (1974), Sebring-Vanguard planned to build "no more than" 2,500 vehicles. These cars, the company claimed, could be operated for 1.5 cents per mile. This a figure that is commonly tossed about in the promotion of EVs, and it has two flaws if compared to the 16 to 18 cents per mile cost of operating a conventional gasoline car. In the first place, the high figure for the conventional car includes everything—licenses, tires, a battery every three years, insurance, maintenance, *and the cost of the vehicle itself*—in addition to the cost of the gasoline needed to fuel the car. Gasoline alone runs about five cents per mile nowadays (ten miles per gallon at fifty cents per gallon) for a standard-size car.

The other flaw in the "fuel" figure used to promote EVs is the fact that the batteries themselves are part of the fuel—

they are partially consumed every time they are charged or discharged. A good set of batteries *perfectly maintained* has a life of about 500 cycles (complete discharges and corresponding recharges), so that a set must be replaced every two or three years depending on mileage. On a small car a set of batteries would cost about $500. Figuring thirty miles to a cycle, one would obtain about fifteen thousand miles from a set of batteries, at a cost of about 3.3 cents per mile *for battery consumption alone*. Add that to the cost of charging the batteries and the total of 4.8 cents per mile comes close to the 5 cents per mile cost of gasoline on a standard-size car. A gasoline car the size and weight of the little EVs on the market today can be operated for one-half the fuel cost of a standard-size car.

The EV has another problem: Due to the phenomenon of self-discharge, all electric storage batteries must be slow-charged at least once a week or trickle-charged constantly when not in use in order to protect the lead plates from excessive sulphation. So an infrequently used EV would incur a small but necessary charging expense even when not in use.

The point is that barring a breakthrough in battery technology there's not going to be any special economy in EV use. The environmental advantages are, however, great and of increasing importance.

So the search for a viable EV goes on. A spokesman for Chicago's Commonwealth Edison Company told a March 1974 Society of Automotive Engineers meeting in Detroit that his company had the capacity to recharge half a million electric automobiles each night. At the time he made the statement there were exactly fifty-one electric cars registered in the State of Illinois. At the same meeting a spokesman for Energy Development Associates of Madison Heights, Michigan, said that the federal government had dropped a seven-year program to study the electric car after just one year and that no government funding was currently being provided for such a project. He mentioned that a zinc-chlorine battery that could extend the range of the electric car to 200 miles may reach prototype stage by 1976 and that it could be in production by 1978.

The prestigious Otis Elevator Company boasts that it is the manufacturer of the world's largest variety of EVs. Moreover, says Otis, it is "in the forefront of the companies that are electrifying the field of personnel and products movement." One type of Otis vehicle, the Electric Delivery Van, was put into use in 1974 by the U.S. Postal Service at Stockton, California. The van has a range of 50 miles at "speeds up to" (not at a constant speed of) 45 mph. It weighs 3,700 pounds and can carry a payload, in addition to the driver, of 500 pounds. Baker Electric, one of the famous old electric-car names, belonged to a company that shifted to the production of industrial warehouse trucks at the start of World War I. It later became a division of Otis Elevator Company, which company with its van, an EV for meter maids, a luxury golf cart, and an acquisition called Electrobus, is on the verge of coming full circle in the electric automobile business. Following the successful Stockton venture, the entire postal service fleet of Cupertino, California, was replaced with thirty Otis vans—the nation's first all-electric fleet.

Electro-Dyne, Inc. of Fort Lauderdale, Florida, began in 1974 a unique method of putting a self-financed EV on the market. The company ran full-page advertisements in several Florida newspapers requesting $100 deposits from future owners of their product. The advance down-money would secure for $3,975, when production starts, a car that will be priced to others at $5,000. Electro-Dyne holds a patent on a new electric motor claimed by the company to be more efficient than anything previously designed. An engineering prototype runs 100 to 125 miles at 35 mph, according to the company, and plans call for improvements in these numbers, which are better than most.

"Our car is different from all others because a local inventor, Eugene Carini, developed a superior motor," said Dick

Potts, president of Electro-Dyne. "In the past, most electric-car makers concentrated on improving the battery."

Electro-Dyne plans to use the deposits it received to facilitate engineering, tooling-up, and general corporate expenses. When the car gets into production it will be 14 feet long, have a 100-inch wheelbase, weigh about 3,000 pounds, and use a large set of standard automobile batteries for its source of fuel-power. Electro-Dyne's four-passenger vehicle will be about the size of a Toyota Corolla. Its batteries will be recharged in the usual way by plugging the vehicle's service cord into a standard household outlet. Charging, says the company, will require only about four hours.

One of the big names in the electric-car field may turn out to be McCulloch Electronics of Los Angeles. That company, in order to have a vehicle of the correct size, weight, and electrodynamic characteristics—and to check on such things as tire scrubbing, tire inflation, steering machinery, etc.—designed a car only for its testing purposes. The company is primarily interested in designing, building, and selling electrical components involved in the production of an electric car. But to show the feasibility and practicality of using its components, the company recognized the importance of setting up its own designed vehicle as a showcase.

The idea is to attract other manufacturers that are interested in building their own vehicles. The company also wants to discourage potential builders who plan merely to purchase components from McCulloch for the purpose of converting an existing vehicle from gasoline to electric power. "You don't wind up with an efficient machine by simply converting a gasoline-engined car to run on batteries," says Robert J. Dunklau, the company's engineering manager. "You have to *design* a vehicle to run on electricity. This includes everything from the ground up. Everything's more important with a low-powered car, from rolling resistance to tire scrubbing." Rolling resistance is the overall total of friction at the wheel bearings and their lubricant, and the necessary flexing of tire sidewalls as well as tire scrubbing, which is a gratuitous friction developed between tire treads and pavement as the tire rolls normally.

The company, a division of McCulloch Corporation, Los Angeles, best known for its line of quality chain saws, learned the hard way about the folly of trying to convert an engined car to a motored one. It found that converted cars did not perform well, and it went to great expense to develop a showcase vehicle for the components it is offering to the market. McCulloch Electronics, since about 1966, has been building batteries and chargers for golf carts. The company also supplied the batteries for the Otis van used by the U.S. Postal Service.

McCulloch offers four basic parts in its system for original-equipment manufacturers: batteries, on-board battery charger, electric motor, and speed controller. The battery pack consists of eighteen 6-volt, lead-acid batteries, each weighing 70 pounds, for a total of 1,260 pounds. Total weight of the vehicle, which has a light fiberglass body, is 2,760 pounds, complete with batteries. The batteries are set side-by-side on a long rack that can be inserted into the front of the vehicle and which runs nearly the entire length of the car from the front bumper to the rear axle. For service, the entire rack can be slid out of the car as a unit.

The battery charger has an output of 25 amperes at 120 volts, and it operates from ordinary house current. The DC motor is rated at 15 horsepower, but it is capable of putting out up to ten times that amount for short periods. The McCulloch speed controller is a solid-state electronic system, the purpose of which is to meter the amount of power needed at any given time. These four basic parts of the system have a total manufacturer's price of about $1,500.

Taking quite the other tack, Electric Fuel Propulsion Corporation of Detroit markets regular production models of Gremlins, Javelins, and Matadors—all American Motors cars —that have been converted from gasoline to electric propulsion. Prices range from about $5,000 on the Gremlin (base-priced at about $2,400 in the conventional gasoline version) to $7,500 on the Javelin and Matador (conventionally, about $3,000). The Gremlin, called the X-144 Electricar after conversion, has a 20-hp motor that is fueled by a 144-volt lead-cobalt battery set that, the manufacturer claims, can give the car an average range of 70 miles at speeds up to 65 mph. The battery set can be recharged in six hours from a 220-volt household air-conditioner-type outlet or in twelve hours from a 120-volt outlet of the type found in every home.

Electric Fuel Propulsion Corp. has been in the business longer than many of its competitors. Robert R. Aronson, president of the firm, described a road-ready car, the Mars II, to a Society of Automotive Engineers meeting in 1968. EFP's X-144 was offered to the first twenty-four employees of Wisconsin Electric Power, Milwaukee, to sign up in 1973 on a nothing-down, no-carrying-charge, $100-a-month-for-forty-months basis.

From the beginning it has been an unwritten rule that electric cars present an unusual, often strange, appearance. They have usually been snub-nosed (because they have no engine up front) and are often high for their width and un-streamlined (when actually they need the lowest possible wind resistance obtainable), compared with conventional gasoline-powered cars. Generally, the more acceptable body designs have come from abroad, notably Italy and Japan.

It is interesting, therefore, that the best-looking electric car that has been designed to date, according to many observers, is an American venture named the Sundancer.

Originally dubbed the Mk 15 for its designer, Robert S. McKee of Palatine, Illinois, the Sundancer was developed as a "test bed" electric vehicle to provide "a most favorable operating environment for state-of-the-art lead-acid storage batteries," according to Bob McKee. Shortly after the completion of the Sundancer in two copies (1970), a committee of the Society of Automotive Engineers prepared a standard (now

The first and only two Sundancer electric automobiles in existence in 1974. (COURTESY OF ESB INC.)

known as "SAE J227, Electric Vehicle Test Procedure, SAE Recommended Practice, March 1971") by which the *range* of electric vehicles could be tested. By mid-1973 only one company was known to have made and published tests according to portions of the J227 standard. That company was ESB Inc., maker of Exide and Willard batteries, and the vehicle was the Sundancer, which was designed and built to ESB's order for the purpose of validating the fact that the lead-acid battery is the logical "starting place" in electric-vehicle engineering.

The Sundancer was powered for the tests by twelve 6-volt golf-cart batteries in a unique T-shaped tray that ran virtually the entire length of the vehicle. By the end of 1972 the twin Sundancers had completed more than 10,000 miles of tests, most of which were in accord with paragraphs 4.1 and 4.2 of J227. The batteries provided residential traffic ranges of 70 to 75 miles and 45 to 50 miles in metropolitan traffic. The batteries first used were good for 300 to 400 charge/discharge cycles. Further test and laboratory work by ESB has resulted in new experimental batteries of slightly higher energy capacity and a life of as many as 1500 cycles, a remarkable improvement.

James F. Norberg, director of ESB's engineering pro-

The battery-of-cells energy source for the 72-volt Sundancer electric automobile. (COURTESY OF ESB INC.)

gram, said, following the initial battery successes, that his company would "focus more than ever before on acceleration and high-rate, short-duration power." The indication there is that the most promising short-term potential for today's lead-acid battery lies in urban and suburban applications rather than in intercity transportation.

In addition to the severe problem of battery capacity, which is not expected to be solved in the foreseeable future, the EVs have another important problem, but it's one that can be solved rather easily—speed controls. A simple rheostat like a backstage dimmer control is not practical because the wire-wound device itself uses a great deal of energy and dissipates it in the form of heat. A simple method of obviating the rheostat is the stepped switch, often used on golf carts. With this method, via a series-parallel arrangement of the batteries, more and more battery capacity and voltage can be called upon for hill climbing or for higher cruising speed. But this method allows a limited selection of speeds with a somewhat jerky change from one to another.

Developments in solid-state control have brought about significant new features for EVs: smooth operation throughout the entire speed range, full-time current limit, easy testing and quick maintenance checks, and component interchangeability throughout a broad range of power consumption. Prior to the employment of solid-state devices, speed control (in such vehicles as warehouse forklifts) was handled via variable resistors (some form of rheostat) in series with the battery

The Motor That Could; the Battery That Couldn't

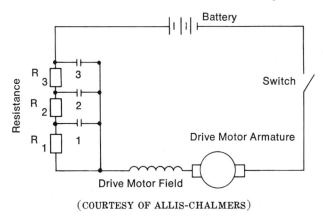

(COURTESY OF ALLIS-CHALMERS)

and some sort of series-wound electric motor. The variation of resistance was accomplished, usually, with a carbon pile (thin sheets of carbon that could be pressed together in a vise-like device for less resistance and loosened for more resistance) or contactor-type control of various types of resistance-windings. Resistors represent power losses; step-type controls, with a number of specific speeds, mean jerky operation.

One of the earlier improvements was a pulse-width-modulation (PWM) control using solid-state devices. Reduction of power losses made possible full-shift operation of lift trucks from one battery charge, and the PWM method also

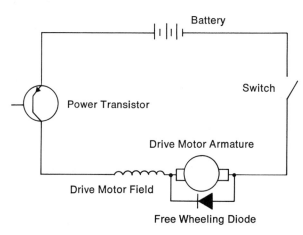

(COURTESY OF ALLIS-CHALMERS)

allows more precise speed control than was possible with older systems. Several parallel-connected germanium transistors are used as a switch that is connected in series with the battery set and drive motor. With this type of switch, a voltage pulse can be applied to the motor a set number of times, usually 120, per second. Variation of vehicle speed is then accomplished merely by changing the width (duration) of the "on-time" of the voltage pulse. As an increase in vehicle speed is called for, the transistor switch is kept "on" for a greater period of time during each pulse. This allows current to flow for a longer time and thus provides more power-per-second to

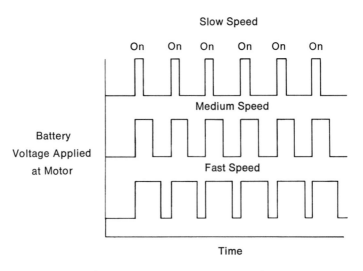

(COURTESY OF ALLIS-CHALMERS)

the motor. Because transistors have low resistance to the flow of electricity, they have very little voltage drop across them; with this smaller loss, a greater percentage of power is available to the drive motor.

Solid-state controls can be utilized to take advantage of a characteristic of the inductive circuit of the drive motor. An inductive circuit tends to retain some current flow when the external voltage source (battery) is disconnected—the current does not stop instantly when the circuit is opened but decays exponentially, the decaying current being called "induced" current. Induced current flowing in the drive motor during the "off-time" of a pulse can be utilized by adding a special "free-wheeling" diode to the control circuit, making a path for the induced current to travel so that it can produce torque. The diode is a one-way switch for DC current, and no current flows through it during on-time of the speed control. But since induced current does flow through it to produce torque during off-time, the average current flowing in a motor under control of a PWM system is higher than average battery

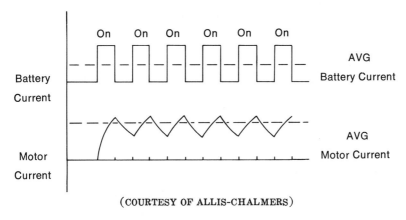

(COURTESY OF ALLIS-CHALMERS)

current. You get something for nothing, in effect, from this type of system.

In a resistor or rheostat type of control system, the greatest losses through heat dissipation occur at the slower speeds when the resistors, rather than the motor, are absorbing the power. With a PWM control, the opposite is true—slower speeds conserve energy. The PWM transistor-type switch allows an almost infinitely variable on-time from zero to full power. This provides delicate control and smooth acceleration. It is easy to provide a selection of current-carrying capacity with a transistor system. Sizing the control is accomplished by adding banks of transistors in, say, 50-ampere groups, so that capacity to match the vehicle's demand can be provided from 200 amps. to 1,000 amps. or more.

Another type of speed control uses a silicon rectifier to vary pulse frequency, rather than duration, to add to or subtract from the amount of power being supplied to the motor. This type, called SCR (silicon controlled rectifier), also acts

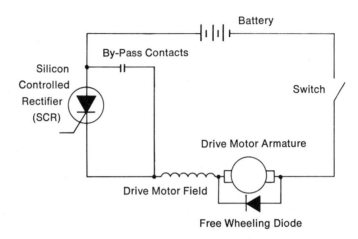

(COURTESY OF ALLIS-CHALMERS)

as a switch controlling the current in the battery-drive motor circuit by changing the frequency of the pulses from about 35 to 300 Hertz (cycles per second). The more frequent the pulses, the more power the motor receives. The free-wheeling diode is also employed to utilize the induced motor current during the off-time of the pulses to provide additional torque.

The SCR system, which is a later development, has the advantage over the PWM system of being able to handle greater amounts of current. But the SCR system has thermal limitations—the silicon must not be called upon to handle high current flow for extended time. This problem is handled through a simple bypass technique—up to 90 percent of top-speed demand can be run through the silicon, beyond which point, with the accelerator pedal fully depressed, full battery voltage is made to bypass the silicon and be applied directly and continuously to the motor. During such periods, the SCR will be cooling because it is out of the circuit.

The Motor That Could; the Battery That Couldn't

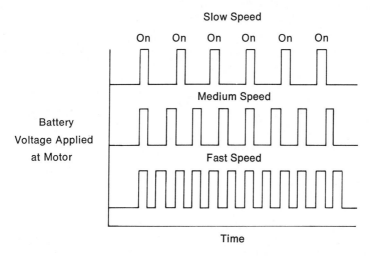

Slow Speed

On   On   On   On   On   On

Medium Speed

Battery
Voltage Applied
at Motor

Fast Speed

Time

(COURTESY OF ALLIS-CHALMERS)

An even later method is much more complicated, but it displays the salutary features of both the PWM and SCR systems while eliminating most of the drawbacks of both. Actually almost a combination of the two older methods, the most up-to-date system utilizes a silicon-transistor integrated circuit. A standard industrial battery of either 36 or 48 volts —the same as used with PWM or SCR—is used with the new system. There are fifteen silicon transistors that govern the amount of current supplied to the motor armature. In this so-called power module each transistor has a diode connected between its base and the drive-current input. Each transistor is thus isolated from the next transistor so that in case of a transistor failure there is no feedback that would turn on other transistors. There is sharing of base current among the transistors because diode resistance increases with any increase of current; this provides an additional margin of base voltage during the off-time of a transistor, to prevent it from turning on. Silicon power transistors of the single diffused type have current-gain characteristics that depend on junction-temperature and collector-current increases. As either junction temperature or collector current becomes greater, current gain decreases rapidly. This protects every transistor from carrying too much current, and paralleling of transistors is allowed without the need for additional current-sharing resistors.

As this book is being written, the research departments of many large companies are working on simplification and other improvements that should cut costs in the face of continued inflation in hardware items; and of course any tendency toward mass production would greatly lower the price to the ultimate consumer.

By far the greatest need in the EV industry is a battery with many times the energy density of the standard lead-acid type. In mid-1974 uncounted scores of research projects large and small were engaged in the search. In late 1973 Gulf and Western Industries, Inc. announced its entry into a joint venture with Occidental Petroleum Corporation to develop a practical zinc-chloride rechargeable-battery system as a power source for metropolitan-area vehicles as well as recreational and industrial cars and trucks. The feasibility of a zinc-chloride battery was demonstrated in 1972 when an experimental system was installed in a compact electric car and driven 150 miles at an average speed of 50 mph on a single charge. Many similar systems have been proposed or are in use, according to David N. Judelson, president of Gulf and Western; but none has been developed with the economic potential to be the power source for a successful urban vehicle.

The zinc chloride battery is made up of sets of zinc and chlorine substrate electrodes arranged in stacks of cells. The cells' faces are filled with aqueous zinc chloride. The battery is equipped with a small plumbing system that includes a circulating pump capable of passing the aqueous zinc chloride from a sump through the cells and back to the sump. The sump is connected with an extra supply of electrolyte and a heat exchanger. The heat exchanger is cooled by refrigeration and is heated by heat-exchange with electrolyte in the sump.

Electrical charging of this type of battery is performed

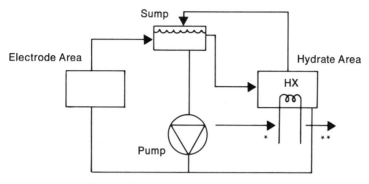

*Heat added by heat exchange during discharge

**Heat removed by refrigeration during charge

Schematic of the zinc-chloride battery with necessary pumping system.

in a way similar to charging of a conventional lead-acid battery; but with the zinc-chloride battery there are additional problems of necessary refrigeration and heat exchange that must be dealt with. These arrangements naturally complicate the system and cause it to be more expensive.

Another important possibility for the future, the sodium-sulfur battery, has been under development at Ford Motor Company since 1963. The work has included not only an investigation into the chemistry involved but also the actual building of engineering prototypes.

Keep in mind that the standard lead-acid battery (the basis of the conventional automobile's electrical system and the on-board source of power for electric golf carts and other electric vehicles) has solid electrodes (plates) of lead and lead dioxide, which are separated by a liquid electrolyte composed of sulfuric acid and water. When this type of battery is discharged, the energy is extracted by an external circuit (motor, lamp, radio) and both electrodes undergo chemical change in the process. The number of cycles of charges and discharges, therefore, is limited, at least in part, by *irreversible* changes that take place in those lead electrodes.

The sodium-sulfur battery, however, has liquid electrodes (liquefied sodium and sulfur). These are separated, by

Auto Engines of Tomorrow

LEAD ACID STORAGE BATTERY   SODIUM SULFUR BATTERY

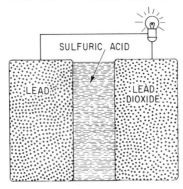
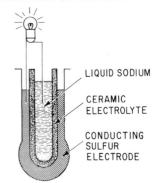

Schematic comparison of conventional lead-acid storage
battery, with liquid electrolyte and solid electrodes, with
sodium-sulfur battery having a solid electrolyte and liquid
electrodes.

contrast, by a *solid* electrolyte, which is a form of ceramic
known as beta-alumina.

When an external circuit is connected and the battery is
discharging, a sodium atom gives up an electron to the external circuit. In doing so, the atom migrates through the
solid electrolyte and reacts with the sulfur on the other side to
form a compound of sodium and sulfur. Ideally, no chemical
or physical change occurs in the ceramic electrolyte during
discharge (or during recharging), and the chemical changes
that took place in the liquid electrodes are completely reversible. Characteristics of the "inert" electrolyte facilitate
important trade-offs between energy and power. In the sodium-
sulfur battery, the total stored energy depends only upon the
total weight of the sodium and sulfur, whereas the power
density (amount of total charge the battery can accept) is in
direct relation to the total surface area of the ceramic electro-
llyte.

The highly attractive potential of the sodium-sulfur battery has been specially explored by Ford since mid-1973 in a
joint venture with the University of Utah and Rensselaer
Polytechnic Institute, under a contract awarded by the National Science Foundation.

The Motor That Could; the Battery That Couldn't

In late 1973 a new type of thermo-electric vehicle motor was tested at one of the laboratories of Volvo, the great Swedish auto maker. The method involves use of a metal alloy that has very good electrical conductivity and poor heat conductivity. Generally the better conductors of electricity (copper, aluminum, silver, gold, etc.) are also good conductors of heat. The Volvo "hybrid" system uses an engine that converts heat (derived from hydrocarbon fuel) into electricity, which is then fed to small electric motors that drive the car.

University of Illinois engineering students at Urbana successfully tested another type of hybrid, or double-engine, vehicle in 1974 that was designed to reduce air pollution and at the same time save on fuel consumption. A project of about a hundred students since 1970, the car requires further refinement before it can be put into commercial production. The experiment, which provides about thirty miles per gallon in urban driving—roughly twice the mileage of a comparable conventional compact—involves a unique linking of internal-combustion and electrical power. A battery-powered electric motor operates at varying speeds and loads to drive the wheels through a gear train. Also under the hood are an electric generator, a motorcycle engine, and associated parts. A large set of batteries is located in the car's rear seat section; if the car goes into production, however, smaller batteries will be used and they will be placed at other locations in the vehicle.

While the driving force to the wheels is the electric motor, the original source of energy comes from the gasoline engine, which in operation has a set and unvarying crankshaft-rotation speed. Under such conditions the engine's exhaust emissions can be kept very clean—it is in stopping and starting, accelerating and decelerating that the gasoline engine is most guilty of polluting. When the Illinois vehicle is decelerating or is stopped, the constant speed engine can continue to charge the batteries. Then, in conditions of start-up, acceleration, and high cruising speed—when the motor demands more than the generator can put out—the batteries can be drawn upon, then recharged whenever motor demand is lower. The present vehicle has a maximum speed of 70 mph, and it can cruise almost indefinitely at 55 mph. Cost of the car is estimated to be 20 to 25 percent higher than a comparable conventional model, but the owner would make up the difference in gasoline savings over a period of about three years. The car's design also allows the use of propane gas instead of gasoline.

In late 1973, the General Services Administration, which is the purchasing arm of the federal government, displayed what it considered the road-ready electric vehicles of the year. There were only five: the Sebring-Vanguard; a T/3 and a T/3 van made by Electronmotion, Inc., Bedford, Massachusetts; a Town and Fairway by Cushman Motors of Lincoln, Nebraska; and a Caroche, made by Club Car, Inc. of Augusta, Georgia. GSA, which buys about 13,000 automobiles a year, says it continuously explores ways of using smaller, cleaner, and more economical-to-operate vehicles.

Any discussion of electric vehicles should give passing mention to the electric motor-flywheel mode and the fuel cell. The flywheel vehicle, already successfully used in Europe, has the capability of taking on electric power via a trolley and using its electric motor to spin a heavy flywheel. Through a clutch arrangement or via a generator and motor-wheels combination, the vehicle, usually a bus, can use the energy stored in the spinning flywheel to propel itself for several city blocks, until it reaches another source of electricity. The vehicle is in effect a trolley bus that doesn't need to be constantly connected to overhead wires. Lockheed Missiles & Space Company has designed a hybrid flywheel system that will allow regular San Francisco trolley buses with a flywheel housed under the lounge seat at the rear of the car to leave their overhead wires where they end at 48th Avenue and Balboa Street, travel to Point Lobos Avenue, then to the trolley wire loop at 33rd Avenue and Geary Boulevard. They would travel 1.5 miles—including a half-mile at a 10 percent grade—on energy stored in their flywheels. While in certain limited applications the flywheel has already proved practicable in large transit vehicles, not much flywheel work has been done at the passenger-car level, primarily for lack of an electrical-source system such as that used by transit companies.

Another example of transit-company capability in the flywheel field is the commendable experiment begun in mid-1974 by New York's Metropolitan Transportation Authority. Subway trains use enormous amounts of electricity, much of which is wastefully converted to heat when the cars are braked. A computer-programmed investigation disclosed that about ten million dollars worth of the annual thirty-three-million-dollar electric bill for the New York subways was wasted in heat due to braking, one of the reasons why subway tunnel air and platforms are so uncomfortably hot in the summertime. In the new experiment, a pair of subway cars were equipped with traction motors that could be automatically converted to generators. This is done via electrical switching whenever the engineer desires to slow down the train. The generators are connected through a control system to other motor/generators that are acting as motors. These motors drive flywheels that weigh 523 pounds; they are accelerated to a speed of 20,000 rpm and are spinning inside vacuum chambers. Thus the decelerating momentum of the train is used to spin the flywheels.

When it is time for the train to start up again, the system is electrically reversed, and the momentum of the spinning flywheel powers the acceleration of the train, quietly and with very little waste heat.

The fuel cell is a device for converting hydrogen directly to electricity. Developed with great success to supply the low-energy needs of space-vehicle components such as radios and tiny motors, the fuel cell poses problems for any high-energy use such as a road vehicle. Nevertheless, in 1974 the Institute Français du Petrole announced a plan to equip a Renault 4L with a hydrogen-air fuel cell. The hydrogen, stored in the form of hydrides, would be synthesized from nuclear heat and readily available raw materials such as water. "Hydrogen, like electricity, is a perfectly clean form of energy, since its use does not have any overall chemical effect on the environment," said an official. Presumably, he was talking about electricity derived from hydroelectric dams. The proposed French vehicle would be able to travel *up to* 50 mph with a range of 143 to 348 miles, depending on speed. Curb weight of the vehicle would be 1,630 pounds, about 230 pounds heavier than the gasoline version. Stating that the fuel-cell vehicle should be ideal for urban use, the official mentioned that the system has a much greater low-speed efficiency than a thermal engine such as gasoline or diesel and that it is virtually noiseless in addition to being pollution-free. If the French vehicle is built and road-tested successfully, it will be the first practical fuel-cell car.

There is a growing conviction that the electric car is ready for a comeback in a modern version. There have been ambitious claims; there have been big expenditures by manufacturers, private inventors, and research groups; and there has been one aborted government investigation into the possibilities of an electric car for the near future. But no one has demonstrated an electric car that is competitive with the conventional gasoline burner of today.

The original drawbacks are extant: limited cruising range, limited speed, poor performance, heavy weight, and—above all—high cost compared with its only competitor in the marketplace. The battery is the big stumbling block, although lead-acid batteries with up to four times the usual energy/weight ratio could be produced on a volume basis right now. Battery makers say, however, that a demand sufficient to justify the high cost of going into production of such a battery is "not there yet."

The Flinders design may be able to exert a significant effect on car design with a printed-circuit motor coupled to a hydrostatic transmission, but at best that kind of design is years away from production.

For the electric car, everything is years away; nothing is "there yet." We'll have to wait a long time for a highway car. But in the meantime, how about a little go-to-the-store runabout that will be quiet, pollution-free . . . and expensive?

4

From Submarine to King of the Road

WHEN the mail steamer *Dresden* completed its crossing of the English Channel and docked at Harwich on September 30, 1913, Rudolf Christian Karl Diesel was not aboard. The world's most famous engine designer had got aboard the night before at Antwerp, and it has to be assumed that he went overboard, but there has never been any evidence as to how or why.

Those who promulgate the suicide theory point to the fact that Diesel's family felt that the man took his own life by going over the side; that he had periods of mental instability for which he received medical care; that he had financial reverses; that he had suffered for years from a painful gout condition and had been plagued with devastating migraine headaches.

On the other hand, Diesel left no suicide note, and although a fastidious records keeper with a legally oriented mind, he left no will; if he had contemplated suicide he never announced, in the manner of the typical suicide, any such intention to anyone; he had plans and business commitments covering the next several years and was, at the time of his disappearance, on his way to talk future business with the British Admiralty—hardly the demeanor of a person likely to cut short his own life.

The accident theory has few backers because there was no storm (the crossing was smoother than usual), no broken railing, no witness; and a careful engineer familiar with ocean vessels wouldn't be likely to fall off a channel ferry absent-mindedly.

The assassination theory is a matter of pure speculation. For weeks after Diesel's disappearance, articles appeared in the press of many nations discussing the probability of economic sabotage on the part of a jealous competitor, the "oil trusts," and "enemy agents." All that we really know is that we know not, and Diesel's death may be forever listed as stemming from reason or cause unknown.

It is interesting to ponder the fact, however, that shortly after Diesel's demise the German submarine fleet became the scourge of the seas and precipitated the entry of the United States in World War I; that the British and Americans engaged in very little submarine activity in that war; that Diesel, a Bavarian national born in France, was friendly to France, England, and especially the United States, where he was always well received and financially as well as socially successful; that at the time of his disappearance he was headed for talks with the British Admiralty; that until the advent of the atomic USS Nautilus in the 1950s the diesel engine was the only practical source of power for submarines and numerous other types of warcraft.

The man whose engine revolutionized all types of heavy-duty commercial transport was born to a German mother and a Slavic father, Bavarian nationals, in Paris on March 18, 1858. The young boy's education began in Paris, where he was able to spend much time in the Museum of Arts and Crafts, where Cugnot's steam wagon of 1769 is on display. When the Franco-German war broke out in 1870, the Diesels, like other foreigners, were ordered to leave France. They made their way to London, where the young Diesel resumed his studies. The British Museum and the South Kensington Museum soon became his favorite haunts. Within eight weeks, however, it was possible for his parents to send twelve-year-old Rudolf to school at Augsburg under the care and tutelage of

Professor Christoph Barnickel. Here the boy spent five studious and happy years.

When he graduated from the trade school at Augsburg, Rudolf was the youngest student ever to pass the final examination, and he not only had the highest grades in his class but he also had made the highest marks ever recorded at the school. As a result of his outstanding record, Diesel won two concurrent scholarships from the Bavarian educational director, which provided him with ample means to attend the Munich Polytechnic *Hochschule* (literally translated "high school," but with course-work level equivalent to that in a college or university). Again, Diesel achieved grades on a scholarly level never before attained by a student at the technical university.

One of the most distinguished faculty members there was Professor Carl von Linde, an authority on heat engines, a lecturer on textile machinery, and the founder of modern refrigeration engineering. Diesel was one of von Linde's most dedicated students. It is likely that at an early point in his engineering career Diesel did the basic thinking on the compression-fired engine that was to bear his name. From his studies under von Linde, Diesel certainly acquired a grasp of the theory of heat engines. He concluded that an engine four times as efficient as the steam engine could be made if combustion took place inside the cylinder and through a wide temperature range.

According to his theory, since the temperature reached depends partly on pressure, extremely high pressure must be produced by means of mechanical compression (via the piston moving in the cylinder toward the combustion chamber and squeezing the air in the chamber). Because of the high compression pressure and the heat it would create, the fuel must be injected after compression takes place in order to avoid premature combustion. This was stated in his thesis, "The Theory and Construction of a Rational Heat Engine," which was published in 1893 and translated into English the following year. (The only other paper he is known to have published, "The Creation of the Diesel Engine," appeared at the very end of his career in 1913.)

Diesel's first employment after graduation was at Gebrüder Sulzer Maschinenfabrik in Winterthur, Switzerland, a position obtained through the influence of his mentor, Professor von Linde. The factory employed more than three hundred men, who built steam boilers, tanks, pumps, and several other types of machines, including the Linde refrigeration equipment designed by Herr Professor.

Diesel had a series of promotions and job changes, including the establishment of a new refrigeration plant in Paris. Thus it happened that his first patent was issued by France, the land of his birth, although he was a German citizen. The patent, issued September 24, 1881, covered an "ice in bottles" idea that was quickly followed by one for a "crystal clear" ice-making process for supplying restaurants and bars—popular then as now in Paris—with table ice. Soon Diesel Clear-Ice Machines were well known, but Diesel himself was not getting rich. A French law stipulated that an employe, such as Diesel, must allow financial rewards from inventions to go to his employer, who by this time was Professor von Linde. With great reluctance Diesel resigned a position he greatly enjoyed; but his employer, the aging professor, graciously asked him to stay on as director of the refrigeration plant in Paris with an honorarium of 3,600 francs a year. By his twenty-fifth year, Diesel's income had climbed to 33,000 francs, worth about that many present-day dollars.

By 1889, Diesel had an engine that ran on ammonia—when it ran. His development of the engine, a hoped-for improvement on the steam engine by, among other things, the substitution of ammonia for water as the working fluid, was marred by many accidents, most of which involved leaks of the ammonia, a choking poison. But eventually Diesel completed a working engine. He was authorized to exhibit the Linde ice machines at the Paris Exposition of 1889, and it was thought that he would also show his ammonia engine. But he didn't, and the reason probably was that he feared a leak that could have been disastrous, in a large crowd, from a public-relations standpoint.

Having expressed the desire to his old professor for a return to Germany, Diesel received a franchise to distribute and sell Linde machines in northern and eastern Germany with a guaranteed base draw of 30,000 francs a year. With things going well and the pressures at a minimum, Diesel suffered less than usual from the terrible headaches that had been bothering him since his college years. Reluctantly at first, because he was unused to Prussian ways, Diesel became a Berliner; then he discovered he liked the place. Berlin became the birthplace of the first engine designed by Diesel. But there had been earlier "diesels."

In 1862 a Frenchman named Alphonse Beau de Rochas proposed an internal-combustion engine that was based on virtually the same principles as the 4-cycle diesel of today: a suction stroke of the piston, a high-compression stroke, ignition (by heat of compression) and expansion (combustion) stroke, and return stroke for exhaust of combustion product. This set the pattern for the great engine of Dr. Nickolaus Otto, which could not be patented because it adhered too closely to Beau de Rochas' principles.

George Brayton of Philadelphia did receive a patent in 1872 on an engine of a type that later became known as a semi-diesel. In the Brayton engine, a metering pump sprayed petroleum fuel on a felt-covered metal grating in the cylinder head. A compressor blew air through the grating, the air picked up droplets of fuel and carried them into a combustion chamber, where the fuel charge was ignited by a continuous flame. Although the engine ran smoothly, it was not considered practical and did not become a commercial success.

An Englishman, Herbert Ackroyd-Stuart, was probably the first to use with success the principle of igniting the fuel charge from the hot walls of the combustion chamber. He experimented for two years, was granted a patent in 1888, and built his first engine in 1890. Hornsby and Company worked with Ackroyd-Stuart on further development of his engine to ready it for production, then began manufacturing it in 1894 under the name Hornsby-Ackroyd. It was an immediate commercial success in a highly industrialized nation experiencing an ever-increasing demand for power. The Hornsby-Ackroyd was put into production in the United States in 1895.

In the meantime, during July 1893, Diesel and his associates had begun practical running tests of an engine that was different enough from all previous types to be patentable. There is some argument as to whether the first fuel was coal dust injected by air pressure or a type of petroleum. In any case it is known that the first petroleum they tried was a dark brown substance too thick to pump easily through the fuel pipes supplying the engine. Later they used a mixture of gasoline and lamp oil before adopting the now-familiar petroleum called diesel oil, which is a cut from crude oil below gasoline and kerosene.

The engine, on which Diesel had received a patent in Berlin in 1892, was an upright single-cylinder machine, with the piston traveling an up-and-down path. The huge cylinder had a bore of 150 millimeters (almost 6 inches) and a 400-mm. stroke (15.75 inches). In this early engine, which was made of cast steel and cast iron, there was no provision for cooling. It developed a pressure of at least 80 atmospheres (a gauge reading of about 1,160 pounds per square inch) before the indicator blew up, sending a shower of glass and metal fragments very near the heads of Diesel and one of his assistants. Tests were resumed the following day and for more than thirty days thereafter, during which time the engine subjected its observers to a series of explosions that covered them, itself, and the shop with a thick layer of soot.

Many modifications, including the addition of a liquid

(water) cooling system, were made on the originally patented version of Diesel's engine. Within a few years the design of Rudolf Diesel became the world standard for what we now know as the diesel engine. All the original tests and early manufacturing, incidentally, were done in the factory of Maschinenfabrik Augsburg, which later became the great Maschinenfabrik Augsburg-Nürnberg, for many years called simply MAN.

Like any enthusiastic innovator, Diesel foresaw a great future for his engine (it became the predominant source of industrial power throughout the world for units up to 5,000 hp), and he felt that America was the place with the greatest potential for the diesel engine. He was right about that, too: Most of the diesels in the world carry a "Made in U.S.A." tag. Yet oddly, none of the three great automotive names found on diesel-powered passenger cars of today is American.

Diesel's later years until his tragic death were repeated episodes of remarkable technical success. The first diesel to be built in the United States was undertaken in 1898 by the Busch-Sulzer Brothers Diesel Engine Company. The president was Adolphus Busch, the Budweiser brewer in St. Louis, who paid Rudolf one million marks in gold, worth $238,000, for North American manufacturing rights. The engine, a 2-cylinder, 60-hp model built on A-frames, was installed to drive an electric generator at the Anheuser-Busch brewery on the St. Louis riverfront.

Although Diesel was right in foretelling that his great engine would reach maturity in America, in his wildest dreams he could not have imagined that his engine would dominate uninterruptedly in almost every location where power was needed in large volume, and without exception wherever coal or water was scarce. Two great wars naturally had the effect of intensifying research and refining production methods, in Germany first when its submarines dominated the seas in the early years of World War I, and in many nations during

89

World War II when world shipping was browbeaten by the Third Reich's diesel-powered U-boats until they were eliminated by diesel-powered sub-chasers of the United States fleet.

Although the diesel has been adapted to satisfy almost every type of power requirement known to man, the big volume of engines today is in commercial highway transport, a mode for which the diesel is almost ideal. But some of the salutary features of the modern diesel that make it best suited to heavy-duty use—such as overdesign in bulk and weight for virtually trouble-free dependability, economy of maintenance, and long life—are not salutary features in passenger cars. And some of its negative aspects—such as noise and hard starting, which are not severe problems in a truck fleet—are deadly negatives in the passenger-car business.

General Motors is America's largest producer of diesel engines—yet it makes none, in the United States, for automobiles. The Opel diesel passenger car, made by GM in Europe, can probably meet the original 1975 emissions standards for hydrocarbons, carbon monoxide, and nitrogen oxides (NOx) with no added emissions controls, GM announced in mid-1973. However, said the company, drastic control measures would be required even to approach the 1976 NOx standard; and in addition, the company expressed the fear that "the already poor acceptability of the diesel passenger car would suffer by such measures." Continuing its pessimistic appraisal of the diesel engine for automobiles, the company added: "Particulates, odor, and noise emissions from passenger-car diesels, while not subject to federal regulation at this time, and not of special concern with the small numbers of such vehicles in use now, could represent a serious environmental concern if a substantial portion of our passenger-car population were converted."

The company did not say, "If the public wants a diesel, with our vast resources we'll clean it up and quiet it down and make it worth the money, as a public service to help the ecology, to help alleviate the gasoline shortage, and as a sound business measure to benefit our stockholders." What the company did say: "General Motors continues to study the diesel engine as a possible future passenger-car engine for some part of the U.S. market, with our R&D [research and development] aimed at reducing its disadvantages."

Exactly what are the disadvantages of the diesel? It depends on your attitude about a number of things, such as noise. The diesel is inherently a noisy engine because of the heavy-duty activities going on inside it—high compression and combustion pressures, the need for slamming big valves open and shut in a hurry, the tendency of the engine towards detonation (knock), and the siren-like noise from the compressor (super-charger). Moreover, partial elimination of another important problem, excessively heavy weight, would make the engine even noiser.

Much of the sound emanating from the diesel's innards cannot be eliminated. But all kinds of excessive sound can be diminished by means of a variety of acoustical techniques. The noise problem could largely be solved by packing the engine in sound-absorbing materials, top, bottom, and sides.

No diesel need smoke. Any diesel that smokes has something wrong with it. Probably the greatest public relations problem the diesel has today is the feeling on the part of many Americans that diesels are inherently dirty. Not so—they can be made to operate much more cleanly than the gasoline engine. Generally speaking, when you see a smoker going down the highway today you're looking at a bus or truck that hasn't had its injectors looked at lately. Injectors are expensive, and when they become worn and allow too much fuel to enter the engine under certain conditions (usually heavy acceleration) it's time to replace them. Naturally, on the part of the independent driver and more particularly on the part of the large competitive operator who needs to squeeze every last mile out of every replaceable part, there's a reluctance to condemn a set of injectors just because they're smoking a bit. So much for the particulates—they're unnecessary.

☙ Odor—especially the pungent, often acrid, stench of the still-volatile hydrocarbon products that usually issue from the stack of a diesel—is not difficult to control. Whether by after-burning to achieve a more complete combustion or by chemical, perhaps catalytic, treatment of the exhaust product, odor can be eliminated. But after all, when you consider that law enforcement is lax regarding simple noise-abatement regulations for trucks, it's unrealistic to expect that the chemical makeup of exhaust gases is going to be very closely inspected. It remains a fact that most of the objectionable aspects of diesel operation are manageable.

Robert Bosch, the great German automotive electrical equipment house, developed in 1927 a fuel injection mechanism that could precisely synchronize the metering of diesel fuel according to the speed of the engine. Until that time the diesel was not practical for passenger-car applications. But with the Bosch invention, the diesel was ready for automobiles. When in 1936 the magnificent Mercedes-Benz 260D became the world's first mass-produced diesel automobile, Rudolf Diesel had been dead for twenty-three years.

The Mercedes-Benz automobile is the world's outstanding example of success with the diesel engine in an automobile. With price being no object, the manufacturer of the luxury car managed to solve most of the problematical objections to diesels in passenger-car applications. In the best diesel car in all the world, noise remains the worst bugaboo. It is susceptible to solution.

In mid-1974 General Motors announced a landmark deal: the great manufacturer would begin making diesel engines in Brazil to be marketed throughout the world. As the plan was put into effect, only commercial and/or heavy-duty engines were scheduled for production at the plant. That is a matter easily changed, however; but that's not the point. What is important is the fact that the United States, although still the world's automotive leader, is gradually continuing to lose its position of preeminence. There are many places in this world

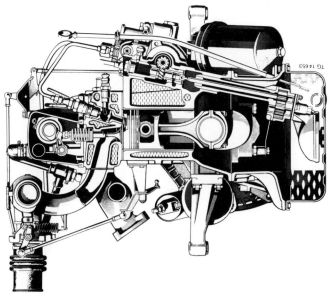

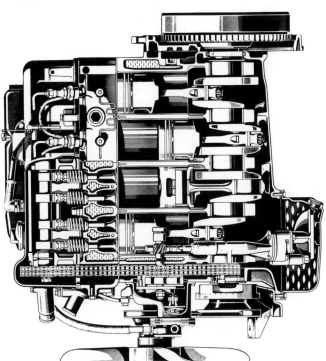

Cutaway views, side and front-end, of Mercedes-Benz 240-D diesel engine. (COURTESY OF MERCEDES-BENZ OF NORTH AMERICA, INC.)

where an engine that makes a bit more noise than others is not taboo; it is to such places that diesel engines will be shipped from Brazil.

In an attempt to "break up America's romance with gasoline autos," Peugeot of France put its "504 Diesel" on the road for a coast-to-coast run in April 1974. At a time when the world was seriously concerned about drastic cutbacks in fuel supplies, the Peugeot scored an impressive 37.3 miles per gallon consumption rate in an eight-day test drive from Los Angeles to Boston. The company hastened to advise the public that the "average" owner could however expect something in the neighborhood of "only" 30 mpg under normal driving conditions. Since the vehicle of modest size is equipped with a fuel-tank capacity of 14.8 gallons, each tankful could carry the car a distance of 400 miles or more. Diesel fuel is more than 20 percent lower in cost than gasoline.

The Peugeot company, which gets rather emotional over the salient features of its 504 Diesel, has issued press releases claiming that, in addition to being more economical to purchase, diesel fuel, when fed into an engine of the same name, is inherently cleaner to burn. The company's reasoning follows this course:

> No special added-on emission control devices are needed to curb pollutants. Since there is no limit on the engine's air intake, and since the engine has no throttle valve, complete combustion of the fuel intake is assured in the diesel. This accounts for the diesel's fuel economy as well as its low emission of carbon monoxide and unburned hydrocarbons. The type of smoke which is emitted from the exhaust when the engine is warming up is not a contributing factor in the formation of smog.

The above is pure public relations talk. PR people rarely are trained in the technical aspects of the products they are selling, and often they deal euphemistically with a subject at hand. "No limit on air intake" is an incorrect way of saying that the driver does not control air intake; "complete" combustion does not occur in the Peugeot, or in any other diesel, or in any other type of engine.

But it is true that the diesel, as a class, does a good job of burning fuel, and diesel fuel is comparatively cheap.

Since Rudolf's first experiment, there have been problems in starting diesel engines. The compression is necessarily so great that the cranking load on an electric motor of the type used as a starter for gasoline engines is often excessive. On heavy-duty engines, such as in highway tractors and bulldozers, there is a compression-release system that can be activated momentarily while the engine is brought up to a fast cranking speed by the starter; then, with the engine spinning rapidly, the release mechanism is deactivated, causing instant compression that usually results in ignition.

In smaller engines, ignition can be encouraged by the use of an electric glow-plug to provide some initial heat inside the combustion chamber. The preheat method is utilized in the Peugeot "by a simple flick of a switch," as the press releases put it. Once a diesel is started it warms up very quickly and tends to run steadily.

The diesel, having no electric ignition system, is without spark plugs, distributor, coil, condenser, points, high tension cables, as well as carburetor, all being parts that require regular maintenance or periodic replacement and that account for nearly all the under-hood maintenance expense of gasoline automobiles. There is a periodic maintenance requirement on the fuel injection system, but that is not a big cost-per-mile item. Diesels should have their lubricating oil changed rather frequently (every 1,500 miles on a Peugeot) and periodic checks of the fuel, air, oil, and coolant filtering systems are called for. That's about all the engine maintenance there is.

The Peugeot 504 sedan, at $5,800 in 1974, was several thousand dollars lower in cost than the Mercedes-Benz, which was the only other diesel passenger car available in the United States during the mid-1970s. No American manufacturer was willing to disclose plans, if any, for marketing a diesel car in the foreseeable future. Peugeot, which claims to be the oldest and largest diesel passenger car manufacturer in the world, first installed a diesel engine in an automobile in 1923. Between 1965 and 1975 the company built more than half a million diesels of the type it now markets, an in-line, 4-cylinder model with Bosch fuel injection. But few have been sold in the United States, where the market for diesel autos has always been next to nil. The Peugeot diesel is well received, however, in Europe, Asia, Africa, and South America. In the United States it costs exactly $1,000 more than the gasoline 504 sedan.

In the mid-1950s there were several thousand diesel taxicabs in America. Sales lasted about three years and then faded because drivers didn't like them, gasoline was cheap, and the small diesel found itself unable to compete with the lower cost of a gasoline engine. But in 1974 the former manufacturer of those taxi diesels, Perkins of England, announced jointly with White Motor Corporation, the American truck manufacturer, a deal that had great significance for the diesel's future in American automobiles.

Sir Monty Pritchard, chairman and managing director of the Perkins Engine Group and group vice-president of the parent company, Massey-Ferguson, Limited, announced that the new joint venture would cause the first foreign engine maker to become involved in diesel-engine manufacture in the United States. Perkins and White agreed to form a new company to build White-designed, in-line, 6-cylinder, 429 cu. in. diesel truck engines at an automated Canton, Ohio, factory. The plans also included introduction in 1977 of two new V-8 diesels of 707 and (with many interchangeable parts) 950 cu. in. displacement, the latter being approximately twice the size of a Cadillac engine. Both the 6-cylinder and the V-8 engines will also be adaptable for big farm machinery. Plans called for the majority of the engines to be exported to Europe.

However, Perkins announced the intention simultaneously to make a strong bid for what the company considers a renewed market for small diesels in the United States. It will manufacture its famous line of passenger-car-size diesels at its Peterborough, England, plant and export them to the United States and Canada for use in city-delivery and taxi fleets. Pritchard credited S. E. Knudsen, White's president, with bringing the deal to fruition. Pritchard said:

> It was expedited by the fuel shortage, which highlights the diesel's advantages even more. We confidently expect that city delivery trucks rapidly will return to diesel power. Perkins once dominated the city delivery market in the U.S. until gasoline engines took over. Now all factors, including a vast improvement in the characteristics of the diesel, point to a rapid return.

He added that Perkins was looking for a U.S. automaker to come up with a taxi design that can be kept static for a number of years. "Otherwise," he said, "suspension and other modifications for new models make the effort unrewarding. Back in the mid-fifties we had 3,000 Plymouths with diesels used as taxis but we felt we were chasing our tails because of the annual model change." But then he brought up an even more telling problem when he recalled, "American hack drivers also were not thrilled by the performance of our 65-hp, 4-cylinder model. They claimed it had no zip." He hastened to add that Perkins currently has diesels in its line that are more suitable for taxi use than the old 65-hp engine.

For heavy-duty applications, the great world-standard over-the-road engine is the Cummins. Available in an extremely wide variety of sizes, the Cummins can be found under the hood of almost every brand of American truck, tractor, road grader, bulldozer, power shovel, dirt handler, and other heavy commercial vehicle. Although the company has announced no plans to produce an engine suitable for a passenger car, its position of leadership in diesel production makes it worth watching, because it's reasonable to expect the first successful American diesel engine for a passenger car to be a spin-off of a heavy-duty version. That's where the experience is.

In early 1974 the Cummins Engine Company of Columbus, Indiana, introduced its new "K" family of engines for meeting increasing line-haul power needs in the 1970s and perhaps the 1980s. The first powerplant announced was the all-new KT-450, rated at 450 hp at 2,000 rpm, with a peak torque of 1,350 pounds-feet at a low 1,500 rpm. The outstanding feature of the KT-450 was its compact size. "While it generates 100 more horsepower and has a 34.5 percent greater displacement, it fits into the same space as an [earlier model] NTC-350," said a company spokesman. All major truck manufacturers had engineered a chassis for the new engine before it was introduced. After more than 60,000 hours of lab testing —the equivalent of 2.5 million miles—tests showed that the new engine would use about 5 percent less fuel than other diesels in the same horsepower range and that it would produce lower emissions.

Scheduled for three million miles of test-truck endurance testing, the engine featured "unusual skewed valve configuration to improve the air swirl and breathing, and reduce [fuel] pumping losses," according to a company official. "The large-diameter [three-inch] high-position camshaft improves the engine's overspeed capability, keeps stress level well within acceptable limits, and permits an extremely short fuel-injection period," he added, continuing, "These features combine to produce a significant improvement in fuel consumption and a basis for meeting 1975 emission standards." Other refinements on the engine that were designed to reduce maintenance costs included almost complete elimination of belts and external fluid connections, use of plug-in accessories, increased filter capacities, and exceptionally high-capacity built-in coolers for the lubricating oil.

The company estimated that the family of engines, when completely filled out, would number about six with a common bore and stroke designed to achieve 100 hp per cylinder with a range of 200 to 1,600 hp. An unusual 12-cylinder engine also was expected to be put into production soon, and there are "other possible configurations" under consideration or actual development; these will or will not be introduced, depending on future market demand. Perhaps the most remarkable feature of Cummins' new "K" series is that it out-diesels its brothers on an aspect for which they are famous: durability. The "K" engines will operate for 500,000 miles or 12,000 hours between overhauls.

The buses of the Chicago Transit Authority seem to travel bumper-to-bumper during rush hours, but the bumpers could never touch, covered as they are with a thick scum from each other's exhaust. Poorly maintained by indifferent officials, CTA buses pour out quantities of smoke that are unrivalled by those of any other city. And they do stink. The fact that the same types of vehicles in service in other cities don't smoke or smell as much serves only to characterize the personnel involved in the very different operations.

Here the point is that the diesel engine is inherently a producer of a smelly exhaust, and that if the diesel is ever to be packed into passenger cars and jammed into rush-hour traffic by the hundreds of thousands, its emissions first will have to be cleaned up. There is virtual assurance from the scientific community (the automotive community has shown less interest) that such a feat can be accomplished, and without much difficulty. In all fairness it must be borne in mind that almost any kind of "cleaning up" in our present technology means a loss of fuel economy. But the gasoline engines suffered that fate and survived (for a while, at least), and it's getting close to the diesel's turn. With the diesel, the job may be easier, although Detroit makes it sound like a task comparable to opening up the Red Sea again.

An aid that came upon the scene in early 1974 may speed the task. It was announced as a replacement for the human nose in unsavory tasks such as sniffing diesel exhaust products to evaluate them qualitatively and, most importantly, quantitatively. It's a chemical analysis unit developed at Arthur D. Little, Inc., Cambridge, Massachusetts, for the purpose of measuring the intensity of odors produced by diesel engines. The first of its kind ever developed, it is expected to stimulate research on ways to reduce and eventually eliminate the odorous compounds in diesel exhaust gases from trucks, buses, locomotives, and off-highway equipment.

To use the device, all a technician has to do is to collect an exhaust sample and extract the odor compounds in a hydrocarbon solvent. With a syringe, the extract is then injected into the unit, where it is automatically divided by a chromatographic column into two separate fractions. One fraction contains the oily-kerosene odor agents in diesel exhaust, and the other has the smoky-burnt agents. Intensity of both fractions is displayed on a strip chart recorder. On the new device, complete analysis of a sample can be performed in about five minutes. The research project, which included development of the measuring device, was sponsored by the Coordinating Research Council, with funds and technical direction provided by the Environmental Protection Agency, the American Petroleum Institute, the Motor Vehicle Manufacturers Association, and the Engine Manufacturers Association.

Specific compounds that cause diesel odors were isolated through liquid chromatography and high-resolution mass spectrometry. There was confirmation of the suspicion that the odorous compounds constituted only a small percentage of the total exhaust volume, ranging from a few parts per million to less than ten parts per billion.

It was discovered that the oily-kerosene odor came from aromatic portions of the fuel not burned during combustion, and that the smoky-burnt odor emanated from oxygenated but incompletely burned aromatics and paraffins in the exhaust gases. Just finding out what it is rates as a partial victory in the war against diesel-exhaust odor. It is worth considering that many more steps in this direction will be taken in the immediate months ahead, because the malodorous exhaust is today one of the chief negative marks against the diesel and one that would become overwhelming if uncontrolled as diesel passenger-car applications increase in future years.

But will the number of diesel cars increase to any significant degree? Obviously, a gasoline shortage would be a boon to the diesel. But in the nation's first gasoline shortage panic, in the fall-winter of 1973–74, there was a concurrent shortage of diesel fuel that created severe problems for the over-the-road trucking industry, which is virtually 100 percent diesel-powered. At times feeling ran so high that many fuel retailers tried to avoid being seen selling diesel fuel to owners of Mercedes-Benz automobiles for fear of resentment on the part of truck-driver customers.

When and if future shortages occur, it may be that gasoline will be short when diesel fuel is not, and perhaps vice versa. To protect the economy and as a matter of good politics, every effort should be made in the future by government and the oil industry to prevent shortages of diesel fuel. In any case, it's almost bound to be cheaper than gasoline because it's easier to refine, and cheaper to use because its greater British thermal unit (BTU) content packs in more miles per gallon. Finally, the diesel engine itself is more efficient—it burns fuel and converts its heat to mechanical force better—than the gasoline engine.

It is difficult to find an automotive authority who can give an objective opinion on anything in his line of business. Every expert worth his salt has irons in the fire in this, the most competitive of all businesses. Nevertheless it is possible to find experts not locked into diesel who will admit that of all the known alternative engines, the diesel has the best chance of substituting, at least in part, for the conventional gasoline engine. The way it's usually put, the diesel has the best chance of soon being *first in line* among the alternatives, because diesel technology is almost complete. That cannot be said of any of the other alternatives except the Wankel, which also is already on the road in many hundreds of thousands of cars.

The practical diesel, however, is about three-quarters of a century older than the practical Wankel. That much more experience could count heavily in favor of the diesel as the race for the market surges closer to the wire.

5

The Wankel Wrangle

ON A FUTURE day, someone is bound to attempt to attract attention by saying that Felix Wankel did not invent the Wankel engine. Against a certain measure it can be certainly shown that Otto did not invent the Otto engine in 1878 (he couldn't get a patent on it because it was too much like the work of the Frenchman, Beau de Rochas, who had already obtained a patent sixteen years earlier). And of course Hitler tried to destroy Viennese booklets published until 1938 that said that the erratic genius Siegfried Marcus (a Jew) invented the world's first gasoline automobile. The Nazis were sold on Gottlieb Daimler and Carl Benz (names whose initials appeared on great engine models like the DB601 that powered the famous Messerschmitt 109 of World War II.

The problem is that we don't have an absolute definition of the word *invent*. Is the inventor the person who first has the idea that such a thing is possible? The person who works with it for years and accomplishes more toward its perfection than anyone else? The person who readies it for the marketplace? Or the person (or by this time, perhaps a committee) who causes it be become in fact a marketing success? Often today an invention is the work of many people in concert, several individuals not connected with each other, or a reverse pyramid in which one person starts a project that is finished by a group.

Auto Engines of Tomorrow

Thomas Alva Edison is recognized without question as the inventor of the incandescent electric light because he saw the need for one, theorized that a current could be forced through a resistive material and cause such molecular friction as to heat the material to the point of incandescence, and then found the material.

By the same standards the world recognizes that Goodyear invented the vulcanization of rubber, Whitney the cotton gin, Burroughs the adding machine, Howe the sewing machine, McCormick the reaper, and so on. With something a hundred times as complicated, such as an internal-combustion engine, assessment of rights and honors is more difficult.

Dr. Wankel was only one of a large number of theoreticians, over the first three-quarters of the twentieth century, who were interested in the idea of a rotary engine. It was a fascinating idea to all who were deeply involved with the reciprocating (conventional) engine and aware of its many shortcomings, chiefly those shortcomings connected with reciprocation itself.

In the engine used in nearly all automobiles today, cylindrical pistons are fitted inside cylinders in such a way as to slide back and forth with each stroke (intake, compression, power, and exhaust) of the four-stroke cycle. In a hand-made or extremely high-quality engine the moving parts can be almost perfectly balanced so that the engine runs quite smoothly. But there is the inescapable fact that nearly all the moving parts of the engine must experience episodes of starting, accelerating to a high speed, rapid decelerating, coming to a complete stop, and starting again as they perform their functions. Such reciprocating actions, no matter how well balanced, create sound, vibration, and wear; and they represent changes of direction of mass that are wasteful.

The stop-start actions that are characteristic of the conventional engine are not confined to the pistons and associated parts. The entire valve train, with two valves for each cylinder, must produce the jump-open, slam-shut motion that admits the

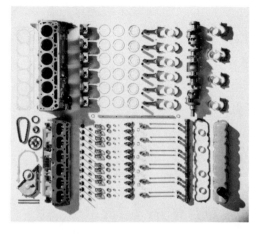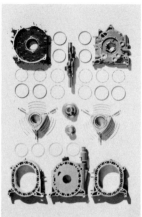

Dramatic evidence of the simplicity and potentially lower production cost of the rotary engine, compared to a conventional six-cylinder reciprocating engine, shows in these parts layouts of the two types of powerplants. (COURTESY OF MAZDA MOTORS OF AMERICA)

fuel charges and exhausts them to the atmosphere. That reciprocating action begins at the camshaft, which converts rotary-shaft motion to lineal motion through the valve lifters, push rods, rocker arms, and finally to the valves themselves, in the typical valve-in-head engine. The valve train produces some of the vibration and much of the noise that emanates from a conventional engine.

The Wankel engine, on the other hand, features moving parts that travel unidirectionally in a circle, hence the term *rotary* that is popularly applied to it. In the Wankel there are no valves in the customary sense, valving being accomplished when the rotor passes over ports that either admit the fuel charge or allow it to be exhausted. More on this later.

With all major-parts movement being rotational and always in the same direction, balancing of a Wankel engine becomes highly simplified, almost as minor a task, in fact, as balancing the armature of an electric motor. In the development of the rotary engine, elimination of *all* reciprocating motions inside it was the goal most inventors aimed at. And although there were attempts to compromise here and there, it has become clear that the pure-rotary thinkers were more correct than the others.

During the five-decade span that Wankel was interested in the rotary and was devoting his time, on and off as he could, to its perfection, a number of other inventors were pursuing the same goal, but via different paths. All the others were wrong, and Wankel was right.

The son of a forest commissioner, Felix Wankel was born on August 13, 1902 in a Black Forest region of southwestern Germany called Swabia. His father, Rudolf, was killed early in World War I when Felix was twelve years old. At graduation from high school, Felix found himself at age nineteen confronted by his country's devastating postwar inflation. Financially unable to continue his studies on a formal basis, the young man found employment in Heidelberg with a technical book publisher. By the time he was twenty-two, however, he was proprietor of his own workshop in Heidelberg. Although the services he provided to customers at the outset of his mechanical career consisted mostly of prosaic operations such as grinding brake drums and other relatively simple tasks, Wankel was developing an appreciation of and a skill in performing the more precise and complex machine-tool work needed in his upcoming years as an inventor and designer.

He was able to satisfy his yearning for further technical education by going to night schools and taking correspondence courses, and early in his mechanical career, probably by 1924, he had begun to study the possibilities of the rotary engine that someday would bear his name. It should be pointed out that Wankel did not invent the first or best rotary engine. He did, however, invent the basic designs that led to the eventual development of the first *successful* rotary engine. For that he rates an important place in automotive history. Development of his basic designs was so rapid that the most successful rotary on the road at the beginning of 1975 resembled Wankel's designs only on a first-cousin basis.

The rotary idea is not new. An Italian, Agostino Ramelli, built a rotary water pump in 1588, and James Watt proposed a rotary steam engine two hundred years later. In this century John F. Cooley obtained patents on what is probably the world's first internal-combustion rotary; his patents were dated 1903. By 1911 the Swiss machine-tool company, Oerlikon, was working on the same type of engine. Yet no person and no company made progress of great commercial significance until Wankel took up the challenge; and even he struggled with it for three decades before approaching success.

Early in that long period, Wankel became an expert on compressors, and he invented several types of seals used in compressors and engines. The principles involved in designing engines and compressors are almost identical. Whereas an engine contains an expanding-gas form of energy that it converts to mechanical motion, a compressor receives energy from an outside source and converts air or other gas to a high-pressure condition. In the same way that an electric motor and a generator can have interchangeable characteristics, engines and compressors are similar in appearance and action while performing opposite functions. Therefore a compressor expert would automatically become an engine expert, and a person like Wankel who dealt with the sealing problems of one could be capable of handling the sealing of the other. Since many of the problems in designing a rotary engine were problems of sealing, it was a natural thing for Wankel to concern himself with rotary-engine ideas throughout his long career.

His first patent, dated 1929, was not connected even remotely with a rotary engine, however. It covered a double-acting-piston internal-combustion engine that was similar to certain kinds of steam engines. A reciprocating piston, crowned at both ends, was inside a cylinder that had a combustion chamber at both ends. Wankel was aware that his first patented work was hardly an advancement toward the rotary he had had in mind for several years. In fact, there is evidence that at least once during this period Wankel was convinced that no further work should be invested in the rotary. As a seals expert he knew that great advances would have to be made in sealing techniques before the rotary could be perfected.

Wankel's work with seals continued, and his interest in the rotary engine picked up again. But he made no real progress on any of his engine ideas. Daimler-Benz hired him to conduct research directly on rotary engines and associated valves and seals. He worked at this for about a year and then switched to BMW, where he developed a cylindrical-piston engine with rotary valves. He was successful, receiving a patent shortly for a sealing element Wankel called a "packing body." This was a method of using a small low-friction surface for effective sealing without heavy spring-loading by using the pressure of the combustion gas to set the seal against its mating surface.

By 1936 Wankel had obtained so many sealing patents that he was considered preeminent in seals used in the field of compressors, rotary valves, and internal-combustion engines. Two years earlier he had applied for a patent on his first rotary engine. It was heavy, bulky, complicated, and inefficient, but it proved the worth of many of his ideas in the field of sealing.

Wankel was in a specially productive period of his life in the early 1930s when Adolf Hitler was taking over Germany. It is not clear how the engineer-inventor got involved in politics, but he somehow discovered and gave publicity to an embezzlement scheme perpetrated by the National-Socialist German Workers (Nazi) Party and for his pains got himself thrown into jail by Hitler when the dictator came into power. Wankel was released in 1935 after several months as a political prisoner, and he moved from Heidelberg back to his hometown, Lahr, in Swabia. Hitler's Air Ministry decided it needed him, however, so in less than a year he was called upon and set up in his own technical center, Wankel Versuchswerkstatten (WVW), in Lindau, Germany.

He worked on a rotary-valve aircraft engine for Hermann Goering's Luftwaffe, and he was a consultant for Daimler-Benz during the perfecting of the great DB601 engine used in the Messerschmitt 109 of World War II fame. Wankel also supplied rotary valves to the Junkers company, another aircraft manufacturer, and Germany's surrender in 1945 prevented completion of a scheduled run of a hundred torpedo engines on which Wankel had worked.

French occupation forces dismantled the WVW and imprisoned Wankel for about a year. With the help of some influential friends, however, he was reestablished by 1951 in his own new technical center, the Technische Entwicklungsstelle, located at Lindau am Bodensee, West Germany, not far from the original site of his WVW. From that time on, Wankel followed a direct course leading to the world's first successful rotary-combustion engine, to a great fortune and world-wide fame, and to an honorary doctorate from the Munich Technical Institute in 1969.

The opening of Wankel's new technical center in 1951 was followed by a long series of contracts with NSU, then a prominent German manufacturer of motorcycles. Wankel worked on rotary valves for motorcycle-engine application, and in 1954 he built several highly successful compressors (superchargers) for cycle use. For the first time in those compressors, Wankel used his complete seal grid. His work with compressors excited his interest in engines once again, and this time the inventor made greater progress than ever before.

Wankel's first rotary engine experiment at NSU was dubbed DKM (*Drehkolbenmotor* in German, literally "rotary piston engine"). Tested in 1957, it ran only long enough to obtain a torque reading. That engine's rotor moved concentrically, and its outer housing rotated. Although obviously never designed to be installed in an automobile, the DKM proved rotary feasibility beyond a doubt.

Dr. Walter Froede, an associate of Wankel at NSU, took the rotary at that point and developed from it the KKM series (*Kreiskolbenmotor*, or "circuitous piston engine"). Note that both the DKM and the KKM are called "piston" engines by their inventors, and indeed they are, despite the common practice of identifying the conventional reciprocating engine as a piston engine to discriminate between it and the rotary. The rotor of a Wankel engine is a piston—or if you wish, three pistons in one piece. A piston does not have to be cylindrical in shape, nor does it have to describe a reciprocating motion.

In the KKM design the outer housing remained stationary, making the engine susceptible of being installed in an automobile or truck, and the eccentric movement of the rotor became a standard now identified with all Wankel engines. Dr. Froede's KKM, tested in 1957, weighed only 37.4 pounds with a cast-iron housing and an incredible 23.2 pounds with an aluminum housing. It was the first Wankel engine as we know it today, but it went through several more important stages of development before it was a practical automobile engine. The first-generation Wankel engine, the DKM-54, had 54 cubic centimeters of volume. Dr. Froede's first model was a KKM-125 (125 c.c. in size).

In July 1959 endurance tests were begun on the KKM-250. Successful tests up to 1,000 hours were completed by the end of the year on the 250-c.c. engine, which weighed about 48 pounds and developed a peak horsepower of 44 at 9,000 rpm. In December NSU introduced the KKM-250 to the public, and news of it went around the world.

Officials of Toyo Kogyo Company, Limited, Hiroshima, Japan, a manufacturer of machine tools, motorycycles, three-wheel delivery cycles, subcompact autos, and small trucks, immediately took an interest in the new German engine. Toyo Kogyo had been primarily a light-truck builder right after

World War II but had reentered the passenger-car field in 1960. The company, under Tsuneji Matsuda (pronounced something like "Mazda"), who was then president, attempted to open talks with NSU regarding a license agreement for research, development, and manufacture of the Wankel engine in Japan but was at first curtly rebuffed. A few months later, through the good offices of the West German ambassador, Dr. Wilhelm Haas, who had taken a liking to Toyo Kogyo officials during a visit to Hiroshima, an agreement dated October 12, 1960 was attained.

The first prototype plans from Germany caused the Japanese engineers to produce a very bad engine. The first prototype engine shipped from Germany to the Toyo Kogyo test lab at Hiroshima performed very badly. Both the prototype, a KKM-400, and the Japanese engine built from the German plans showed excessive roughness at idle speeds, emitted clouds of smoke, and consumed oil at a rate "beyond all practical use," in the words of one of the engineers involved. To top off the disappointment, the engine made in Japan lost its power-output capabilities at 200 hours on the bench. The dismantled engine showed "chatter marks"—signs of irregularity in the mating of the apex seals at each point of the triangular rotor with the inner lining of the housing. This meant that the electroplated lining was tearing up. Then began an agonizing search for materials, a frustrating series of half-successful tests, many computer-directed changes, and a few episodes of nearly giving up.

But the Japanese engineers, with a stubbornness that would have made any German inventor proud, stayed right in there. In 1967 they had an engine ready for the road.

Almost ten years before, Curtiss-Wright, the American manufacturer of the "Cyclone" and many other famous aircraft engines, had made a brilliant purchase of the American rights to the Wankel engine for $2 million from NSU Motorenwerke AG. Westinghouse, via a development contract with Curtiss-Wright, produced rotary-powered electricity generation equipment of light weight, mobility, dependability, and comparatively low fuel consumption for military uses.

A sub-license to Outboard Marine Corporation in 1966 resulted in the development of sophisticated Johnson and Evinrude outboard rotary engines for snowmobile and marine use.

In the Wankel engine an internal-combustion process chemically identical to that in the reciprocating engine takes place—an air-fuel mixture is taken in, compressed, ignited, divested of some of its energy, and exhausted—but the mechanical operation is different. In a conventional engine we have a piston traveling inside a cylinder, like the piston-cylinder setup of an insect sprayer. In the Wankel rotary the same internal-combustion process is accomplished but in an entirely different way and with somewhat different results. Instead of round pistons-in-cylinders, rotors-in-housings are used. The rotors are roughly triangular in shape and they turn inside specially shaped raceways that have been machined on the internal surfaces of the housings.

If the rotors were merely to spin inside a cylindrical housing, in the way an electric-motor armature spins inside the field coils, there would be no engine of any worth. The four episodes of the four-stroke cycle would not be satisfactory and work could not be performed efficiently. But the rotor in a rotary engine performs two concurrent movements: it spins like a motor armature and simultaneously travels an eccentric path to follow the configuration of the inside surface of the housing.

The fat-figure-eight shape of the inner surface of the housing (cylinder, if you will) is called an epitrochoid. So the rotor not only spins about its shaft but also simultaneously travels an eccentric path to follow the epitrochoidal configuration of the inside surface of the housing.

The rotary engine begins its intake of the gasoline-air mixture when the rotor passes over a port in the side of the housing at a time when there is only a small-volume area between the rotor and its housing. As the rotor continues to

Interior view of Mazda's two-rotor rotary engine with manual transmission. (COURTESY OF MAZDA MOTORS OF AMERICA)

The heart of the Mazda rotary engine, with end frame removed, showing the rotor's position inside the epitrochoidal chamber. (COURTESY OF MAZDA MOTORS OF AMERICA)

move, the volume of that area increases rapidly as the rotor, while spinning, moves away from the port. This compares to the intake stroke of a conventional piston. The increase in volume of the chamber causes suction that brings the fuel charge into the engine.

After the fuel charge is inducted, the next apex (point) of the rotor covers the intake port, and the spinning-plus-eccentric movement of the rotor begins to decrease the volume of what is about to become the combustion chamber, compressing the fuel mixture. By the time the mixture is fully compressed the portion of the rotor now being considered has moved around the housing's lining to the place where the fuel charge is ignited. At this position on a Mazda rotary there are two spark plugs, a "leading" and a "trailing" plug. The leading plug, the first one encountered by the compressing portion of the rotor, fires to initiate the flame front, and then the trailing plug fires to assist in completion of combustion of this particular fuel charge.

Now the rotor is under the power of the charge we have been following, and as the rotor continues to turn, its also-eccentric movement allows the rapidly expanding combustion gases to push it away from the raceway. As this happens, the volume of the moving combustion chamber grows larger, and when the rotor uncovers the exhaust port the still-combusting charge "pops" through and becomes the leading portion of the exhaust. The following apex of the rotor continues toward the port and clears most of the remaining exhaust products from the engine. By this time the portion of the rotor we have been considering has already become an intake chamber of expanding volume again, and a new cycle for it is under way.

Only one of the three sides of the triangular rotor has been spotlighted in this discussion, during which the other two sides have been performing identical functions in turn. And if, as is usual, the engine is a two-rotor type, the other rotor in its own housing has been functioning in the same order as the first rotor but stepped 60 degrees out of phase with it.

One of the rotary's chief attractions to engineers is its ability to put out more power than a reciprocating engine of comparable size. This means that the rotary can be much smaller than a reciprocating engine of comparable power. Let's see why that is and then examine its peculiar significance as an antipollution feature.

In a conventional four-stroke-cycle engine, each piston must make four complete back-and-forth strokes, involving two revolutions of the crankshaft, for every firing of a fuel charge. In a 2-cylinder conventional engine, there are two firings for two revolutions. There is, however, a firing per revolution of the crankshaft for each rotor of a Wankel engine, so that in a two-rotor rotary two firings occur with one revolution of the crankshaft. The rotary thus produces considerably more power than a conventional piston engine of the same cubic-inch capacity. A conventional engine with the power output capability of the Mazda rotary, for example, would be

1. Intake.	**2. Compression.**	**3. Ignition.**	**4. Exhaust.**
Fuel/air mixture is drawn into combustion chamber by revolving rotor through intake port (upper left). No valves or valve-operating mechanism needed.	As rotor continues revolving, it reduces space in chamber containing fuel and air. This compresses mixture.	Fuel/air mixture now fully compressed. Leading sparkplug fires. A split-second later, following plug fires to assure complete combustion.	Exploding mixture drives rotor, providing power. Rotor then expels gases through exhaust port.

The four episodes—intake, compression, ignition, exhaust— of the rotary engine by Mazda. (COURTESY OF MAZDA MOTORS OF AMERICA)

expected to be a 4-cylinder or even 6-cylinder model with a great many more parts; thus space-saving is another important advantage of the rotary design.

Every internal-combustion engine must be equipped with antipollution devices in order to have an exhaust clean enough to comply with increasingly stringent federal standards regarding emissions. Some of the devices are simply critically adjusted parts that are built into the engine and take up no unusual amount of space; others are add-on gadgets of small size (hoses, connectors, valves, etc.) that can be accommodated under the hood without great difficulty. But as federal and state emission regulations become more severe, space under the hood for larger antipollution parts, such as thermal reactors or catalytic converters, becomes scarce indeed. Nowadays one of the chief arguments in favor of the Wankel engine is its small size that leaves plenty of room in the engine compartment for government-mandated gadgetry.

In addition to the Mazda of Toyo Kogyo, only one other automobile rotary had reached the road-ready stage by mid-1974: the NSU rotary, as mounted in Prinz and Ro 80 sports cars made by NSU. The NSU rotary has not been a resounding success and as of 1974 had not been certified for sale in the United States.

116

Ford Motor Company has said repeatedly that it will not manufacture a Wankel engine, "Not in my lifetime," as Henry Ford II is said to have put it. Chrysler Corporation, through Alan G. Loofbourrow, vice-president in charge of engineering and research, has been bad-mouthing the Wankel for years as "basically a dirty engine," one that does not get good fuel mileage, and an engine that can be "rough as a cob." American Motors has adopted a wait-and-see attitude; it can always buy Wankel engines from another manufacturer if it decides to market a rotary in its line of small cars. The first American-made Wankel was produced in 1972 by Outboard Marine Corporation for a snowmobile and later as a marine outboard powerplant.

The big question-mark is General Motors. The world's largest manufacturer seldom makes big mistakes, and it soon will have $50 million tied up in Wankel license fees. By mid-1974 it reportedly had had a road-ready Wankel (General Motors Rotary Combustion) engine for months but for some reason did not market it, as predicted, at announcement time for the 1975 models. Some speculation points to the fuel short-age and GM's reluctance to offer an engine that does not feature "improved" gas mileage compared with what is al-ready on the market. Other speculation deals with GM's possible inability at this time to build a Wankel engine cheaply enough.

The Mazda Wankel commands a premium of between $500 and $600 over the conventional 4-cylinder Mazda engine. One official statement from Chevrolet (the GM division most likely to begin marketing of a Wankel) that dates back to 1973 indicated that the premium a purchaser would have to pay for a Vega or other small Wankel-powered Chevrolet product might be as high as $1,000. That's a terriffc premium, enough to scare a sales manager.

In 1974, GM, which had previously issued many news releases on its upcoming GMRC engine, had several of its officials talking about it to the press and in public speeches, and had made public displays of prototype models, suddenly put the clamps on information about the GMRC. Whether this was merely a reaction to too much publicity both good and bad, whether GM developed a serious technical or economic problem with the GMRC, or whether the company is simply making ready for a stunning opening announcement of its latest and most glamorous product, no one who is not a high official knows.

Meantime, the highly successful Toyo Kogyo has come out with its RX-4 Mazda-Wankel engine and even offers it now in a pickup truck. The Mazda was a temporary victim of the fuel crisis because it used more gasoline than other cars of its size and weight. The main cause of this problem was an antipollution device called a thermal reactor. The thermal reactor is a muffler-like device that is bolted to the lower right side of the Mazda engine, at the exhaust ports where the exhaust manifold would be installed on a conventional engine. Hot exhaust is delivered by the engine into the thermal reactor. Fresh air is delivered there also, via a pump. The configuration of the thermal reactor causes a swirling action of both the air and the exhaust gases, so that they are intimately blended. The oxygen in the air supports a continuing combustion of the gases expelled from the engine; and the more complete combustion that results greatly diminishes the amounts of noxious emissions in the engine's exhaust.

In order for the thermal reactor to work correctly, it must be supplied not only with fresh air but also with exhaust gases that are rich in unburned hydrocarbons. In order to insure that the thermal reactor is so provided, the Mazda engine is supplied from its carburetor with a rather rich fuel-air mixture (more gasoline and less air than in conventional engines) ; and the ignition is retarded so that combustion begins late and is continuing as the exhaust gas exits the engine with a torch-like blast.

That combination of late spark and rich mixture somewhat degrades the thermal efficiency of the engine and causes it to give a poorer fuel mileage than can be expected from a conventional engine. When unleaded gasoline becomes universally available and a catalytic device can be substituted for the thermal afterburner, the Mazda will operate as economically as any other internal-combustion engine, says Robert Brooks, a Chicago-based industrial management consultant who is known as an authority on the Wankel engine.

The EPA fuel-mileage tests that showed the Mazda to be a fuel guzzler turned out to be erroneous. In 1973 the EPA published a mileage extrapolation from its emission certification tests conducted during the year. Each test involved a cold start and the running of each car 7.5 miles on a dynamometer, a method of roughly simulating road driving by putting a car on rollers. The EPA published a finding that indicated that the little Mazda got only 10.7 mpg, which was 30 percent less than the lowest other vehicle in Mazda's weight class and almost 60 percent less than the best performer.

C. R. Brown, who until July 1974 was general manager of Mazda Motors of America, violently objected to that bad fuel-mileage finding, saying that the Wankel engine cannot be tested fairly with the same procedures that are correct for conventional engines. Mazda did a number of mileage tests that would seem to prove that the Wankel engine did have better fuel mileage than the EPA tests indicated. Moreover, United States Auto Club–sponsored cross-country fuel-mileage tests of five rotary-engine Mazdas at an average speed of 53.5 mph produced an average fuel-mileage figure of 20.6 mpg, in early 1974.

Under extreme pressure from the publicity Brown was obtaining, the EPA consented to retest Mazda models RX-2, RX-3, and RX-4 against a Saab, a Vega with automatic transmission, a Gremlin, a Vega with manual transmission, and a Torino. While the Mazdas bested only the Torino in the new tests, the Wankel-powered cars did manage to average between 17 and 18 mpg in all-around driving and more than 20 mpg in highway driving. The original EPA score of 10.7

mpg in city driving (simulated) was improved in the special EPA test to an average 13.1 mpg.

There has always been another sales bugaboo for Mazda: the remored possibility that, in the oft-abusive hands of the public, the Mazda Wankel might be short-lived. After hundreds of millions of public miles on the road to provide them with the assurance they were hoping for, Mazda officials in early 1974 stunned the American automobile industry by offering a three-year or 50,000-mile engine warranty, something not offered by Detroit, something available from only one other car manufacturer: Rolls-Royce.

In January 1974, *Motor* magazine of Britain reported tests on the NSU Spider with the latest 500-c.c. single-rotor Wankel engine that showed a fuel economy of 36 miles per gallon, making it as good as or better than a Morris Cooper, MG Midget, Lotus Elan, Triumph Spitfire, and the MG-B Roadster.

In early February, 1974, Volkswagen announced its decision to introduce in 1976 a Wankel-powered luxury car in the $7,000 price class. VW said its 2-rotor engine will be mounted in a car patterned after the NSU Ro 80 that is sold in Europe to compete with Mercedes-Benz.

Of all the alternative engines considered in this book, the Wankel and the diesel are the only ones that are "here now." They also are the only ones that can be expected to compete successfully with the conventional auto engine in the immediate future. The Wankel, however, is the only alternative automobile engine in great volume production now, the only one to begin to receive acceptance from the American public.

Whether the big-selling Wankel of the future will carry the label of Mazda, one of the famous names of General Motors, a tag of Volkswagen, NSU, or Mercedes-Benz, or a brand that will come as a big surprise, is anybody's guess in the mid-1970s. It's a safe bet for the foreseeable future, however, that whoever sells Wankel will sell well.

6

With a Whine and a Whoosh

ASKED why the Ford gas turbine Model 707 was put into production when it apparently was not fully engineered, Henry Ford II gave the kind of reply seldom heard in the automotive industry: "We took a risk. We thought we were going to have some breakthroughs, and we didn't get them."

Model 707 represents the abortive attempt of one of the world's great manufacturers to put a turbine-powered truck of large commercial size on the road in massive numbers. It is an excellent example of an industrial giant being temporarily thwarted by technological problems beyond its control. No one should infer from that single miss, however, that the gas turbine is an unsuccessful engine. All one needs to do is unplug his ears and glance skyward in the direction of the next tornado-like sound to see a craft propelled by a propjet or purejet engine (gas turbines both) that is eminently successful.

The gas turbine for aircraft is the simplest of all modern engines. In some ways it is ideal for high-altitude cruising at high speed. It is quite dependable—flameouts and other problems causing engine failure are rare—maintenance is uncomplicated and comparatively inexpensive, the gas turbine is durable and provides extremely long service, and at altitude its fuel-consumption rate is acceptable.

Steam-operated turbines go back to ancient history, and

121

a number of later inventors made attempts to substitute expanding combustion products for steam as a more convenient method of applying power to a turbine wheel. John Barber of England was the most outstanding among those who recognized the possibilities of internal combustion applied to a turbine. His patent of 1791 is the first recorded description of a gas turbine engine.

By the beginning of the twentieth century many experimenters were building gas turbines, but none was successful. During the nineteenth century the reciprocating steam engine was dominant, and it was followed by the steam turbine, which almost completely superseded the reciprocating type. Development of the reciprocating gasoline engine was contemporaneous with the perfection of the steam turbine; but it was roughly half a century later before the gas turbine received any appreciable attention.

The gas turbine required three things before it could be brought into practical use. It needed a high-volume and high-compression-ratio compressor, because its cycle required tremendous quantities of high-pressure working air. The second requirement was a general improvement in the efficiency of the turbine wheel itself, the compressor, and associated air passages. These were basically aerodynamic problems. The third necessity was a high-temperature material for combustion-chamber and turbine-wheel construction. This demanded mid-twentieth-century metallurgy.

As a prospective competitor to the diesel in trucks, and especially as a projected replacement for the spark-fired gasoline engine in automobiles, the gas turbine still has a long way to go. But it is almost bound to make the grade in both applications, possibly by the mid-1980s in autos and probably earlier in trucks.

Unlike Ford, which in 1973 ceased production of its Model 707 and put its Model 710 back into engineering and research for more development, the Detroit Diesel Allison Division of General Motors began stepping up its production of truck turbines in 1973. DDA announced in September that it would produce another 46 engines during the following twelve months, bringing to 129 the number built since the beginning of its Indianapolis pilot production program in 1971. According to the division's schedule, about two-thirds of the 1974 engines were the 450-hp GT 505 model, and the remainder were the much smaller 350-hp GT 404. Most of the 1974 production run were allocated to truck and coach fleets for the purpose of gaining additional fleet experience, but a few units were going into watercraft and industrial operations.

In early 1974, Goldston, Inc., an Eden, North Carolina, truck-leasing firm, was selected by White Motor Corporation's Freightliner Division to conduct the first in-fleet tests of its new Turboliner, which was powered by a DDA GT 404. William D. Goldston, Jr., president of the leasing company that was selected from dozens of applicants, said that the Turboliner would be tested thoroughly in forty-two states over all types of terrain and with varying load and distance factors. He agreed that his company would keep records on fuel consumption related to gross combination weight hauled, maintenance and operational data, and pertinent comments from drivers, shop personnel, and management. "I believe the trucking industry will be entering a new phase in the next five years, led by the development of new power sources," said Goldston, adding, "The gas turbine just might prove to be the 'million mile' engine we've all been looking for."

Indeed it might. And with a few breakthroughs in addition to those so ardently wished for by Henry Ford II, the gas turbine might become the passenger-car engine we've all been looking for. So near and yet so far, the turbine for autos. As a departure from the conventional gasoline engine, only the Wankel has had so much money spent on it for development. And even with a long head start the passenger-car turbine couldn't be put on-stream in a factory before the Wankel was. The reasons are several, and they are formidable.

At Chrysler Corporation, turbine work was begun before World War II when an exploratory engineering survey was conducted. It showed that the turbine had possibilities of being an ideal automobile engine *except* that neither materials nor techniques had been brought to the point where cost and time of intensive research were warranted. Chrysler started studies of completely new concepts in gas turbine design before the war was over, and in 1945 the company received a Navy contract to develop an aircraft turboprop engine. Within four years Chrysler presented the Navy with a turboprop engine possessing fuel economy approaching that of aircraft piston engines.

Continuing on their own, Chrysler research scientists and engineers returned to the pursuit of an automotive turbine. They had experimental models on dynamometers (horsepower-measuring devices) and in test vehicles by the early 1950s. The development work centered around improvement of compressors, regenerators, turbine (wheel) sections, burner controls, gears, and accessories. The challenges: fuel consumption could not exceed that of conventional engines; components had to be highly efficient and small; noise had to be in the "tolerable" range; provision for engine braking was mandatory; and acceleration time-lag had to be reasonable. That original list was lengthened a bit as such things as air conditioning and power steering became necessary components in modern cars.

Several quite different qualities of the gas turbine mandated special requirements, such as the need for holding down the temperature of the exhaust gas, the necessity of developing high-temperature materials that would be readily available, and the need for keeping the engine light, compact, reliable, easy to maintain, and competitive in cost with conventional engines. A tall order, but Chrysler research engineers were convinced that intensive research and a full-scale design and development program were warranted by the potentialities of the passenger-car gas turbine engine.

On March 25, 1954 Chrysler disclosed its development and successful road testing of a 1954 production-model Plymouth sport coupe that was powered by a turbine. The engine was rated at 100 hp at the shaft, and it contained a newly designed heat exchanger called a regenerator. That engine component, now considered essential to acceptable fuel economy, was designed to extract heat (that otherwise would be wasted) from the hot exhaust gases and transfer that energy to the incoming air, thus warming the air and lightening the burner's job of raising the gas temperature. The innovation did two important things, actually: it conserved fuel and it lowered the temperature of the exhaust.

A gas turbine engine is schematically a torch (burner) that blows upon a fan (a set of fan blades collectively called the turbine wheel). The burner provides the expansion gases, which are directed at the blades of the wheel. The burner is supplied by a nozzle that sprays fuel (usually kerosene, but any liquid hydrocarbon—diesel oil, gasoline, alcohol, even brandy or French perfume—will work). Oxygen is supplied to support the combustion by means of a compressor (supercharger) at the front of the engine. The compressor is driven by the turbine via a shaft. In the case of a pure-jet aircraft engine, all thrust comes from the reaction of the engine to the blast of the exhaust as it exits the combustion chamber. In a propjet or in any type of automotive gas turbine engine, the turbine itself is geared to the propeller or (in an automobile) to the wheels via a transmission and conventional drive train. A pure jet is even simpler because no gearing is necessary. The engine contains a single shaft with a air compressor mounted at the front, the turbine wheel at the rear, and one or more combustion chambers between them at the center of the engine. Air is taken in at the front, compressed, sent to the combustion chamber(s), injected with a continuous flow of fuel, which is ignited in a continuous burn. The expanding gases then flow through the turbine wheel, revolving it and therefore turning the compressor that is up front on the same shaft; and then the gases exit at the rear of the engine, providing the thrust that propels the aircraft.

A gas turbine engine for automotive use cannot be a gas-thrust machine for obvious reasons (in traffic your "whoosh" would melt the car behind you), and in fact extraordinary measures must be taken to cool the exhaust and delay its departure from the tailpipe until it is safe for it to enter the atmosphere. Noise of great magnitude that would be tolerable at an airport or high in the sky would clearly be unacceptable in traffic, and a gas turbine is inherently noisy at both ends, at the exhaust where the expanding gases set up a steady roar, and at the intake where the compressor has a siren-like effect on the air it is reducing in volume. For automotive applications, Chrysler, Ford and General Motors, as well as other designers both foreign and domestic, have demonstrated that both intake and exhaust noises, as well as gear-drive and accessory sounds, are manageable.

Like the aircraft jet and propjet, the gas turbine for motor vehicles features a basic simplicity that is in stark contrast to its big competitor, the reciprocating engine. The gas turbine has 80 percent fewer parts than a conventional V-8 engine. It has only one spark plug, used only for starting the engine; it never needs a tune-up in the conventional sense; its oil does not need to be periodically changed; it has no cooling system requiring water and antifreeze; and its electrical system is not only simple but minimal.

In Chrysler's first-generation gas turbine of 1954 there were two shafts. At the front of the first shaft was a set of accessory-drive gears, the compressor impeller, the regenerator, and finally the first-stage turbine wheel, which drove the compressor impeller and the accessories. Immediately behind the first turbine wheel, and on its own shaft, was the second-stage turbine wheel, which supplied power to the transmission and through the transmission's double-stage reduction gearing to the driveshaft for propelling the car.

In 1956 the "Turbine Special" was driven from New York to Los Angeles in four days without incident. Data from that trip were analysed and incorporated in a second-generation engine installed in a 1959 Plymouth and driven from Detroit to New York in December 1958.

The third-generation engine, called the CR2A, had a variable turbine-nozzle mechanism which, in a coast-to-coast test of 3,100 miles against an accompanying conventionally-powered car (1962 Dodge), showed fuel economy consistently better (about 20 mpg) than the conventional engine. That has been the case to this day with most gas turbine engines tested in passenger cars in mostly highway driving. But only in very recent years has the economy in city driving begun to approach conventional engines.

Rover of England and Renault of France have built and tested technically successful gas turbine engines for automobiles. Rover was the world's first auto maker to build and test a gas turbine–powered passenger automobile, which was demonstrated at Silverstone Race Track in England on March 9, 1950. Twelve years later the company brought its original "Jet 1" to the United States accompanied by Rover's latest version of a gas turbine passenger car, the "T.4." Rover, a leading auto maker since 1904, began research and development work on gas turbines in 1940. The company was then associated with Britain's Air Commodore Sir Frank Whittle in his early pioneering experiments with aircraft jet engines. Rover built prototypes of those engines, which featured the first electric starting system and the first straight-through combustion chamber layout from which Rolls Royce jet aircraft engines were developed. In 1945 a team of twenty Rover engineers undertook development of small gas turbine engines intended for automobile application.

So far, neither Rover nor Renault—nor any of Detroit's Big Three—has attempted to market a gas turbine engine in competition with the conventional round-piston type. But Chrysler came close more than once, and we're bound to hear more from that company in the next few years.

Fourth-generation turbine powerplants were placed in fifty specially-built Turbine Cars and turned over to the public by Chrysler for free sample driving. In a custom assembly line the Turbine Cars were turned out at the rate of about one per week until the last of the fifty cars was completed in

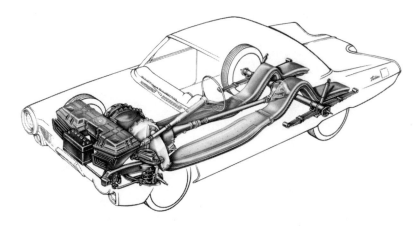

An artist's conception of a phantom view of Chrysler's original Turbine Car, showing important chassis components. Note huge dual exhaust system, needed for both quieting and cooling of exhaust gases. (COURTESY OF CHRYSLER CORP.)

October 1964. The ninety-day trials were continued until January 1966, during which time 203 drivers used forty-six of the fifty vehicles as family cars and accumulated 1,111,330 miles on them, averaging 5,474 miles for each assignee.

Reaction of the drivers to the revolutionary new cars was overwhelmingly enthusiastic (more than thirty-thousand people had applied as test drivers). The smooth, vibration-free operation of the engine was the quality most admired by the users. Next-to-zero maintenance requirement was the second most important advantage in the minds of the "customers," according to the company, which carefully interviewed each driver within two weeks of the conclusion of his or her test period. Another attraction was the Turbine Car's immediate-start capability, especially in extremely cold climates where instant heat in the passenger compartment and no need for antifreeze were also considered strong assets.

In 1967, at a time when many observers were awaiting an "imminent" announcement from Chrysler that the company would introduce a gas-turbine-powered car to the market-place, a surprising contrary announcement was issued: Chrysler did not have a turbine in its immediate future.

George J. Huebner, Jr., Chrysler's director of engineering research, has been "the man behind the turbine" there for more than twenty years. In late 1973, at a meeting of the International Motor Press Association in New York, Huebner revealed why production of a Chrysler turbine for passenger cars was indefinitely postponed six years earlier.

Huebner said it was "a technical decision at that time" (in other words, not a marketing decision). He said that his company had planned to offer the public a turbine-powered car in 1967 but cancelled the plan "when problems with oxides of nitrogen (NOx) began to become apparent." He recalled that the fourth-generation engines sampled by the public were unacceptable in fuel consumption and had a gas generator lag that appeared as balky acceleration to many of the drivers, a high noise level (a high-pitched whine) at idle (although generally quieter at highway speeds than a reciprocating engine), a serious lack of engine retardation for partial braking purposes, and no way to hook up an air conditioner.

A fifth-generation engine showed improvement of most of those old problems, Huebner said, and the current (1973) sixth-generation model had eliminated or substantially solved all the former bugaboos. But new ones had shown up, this time in refinement of the mass-production methods, including volume tooling, needed for turbine components. Here was Huebner's assessment of the situation:

> If gas-turbine engines are to make a significant contribution to clean atmosphere, they must be used in large numbers. So most studies have been made for high-volume production. Here is the heart of the problem: Gas-turbine engines have never been produced in automotive quantities by anyone. To do so will require replacement of current manufacturing techniques with techniques never previously used in large-volume manufacturing. These techniques do exist and many of them are used for moderate-volume pro-

Here are six generations of Chrysler gas turbine engines. The auto is one of fifty built by Chrysler for public experimental use. It mounted the fourth-generation engine displayed at its right front corner. (COURTESY OF CHRYSLER CORP.)

duction, but the labor content is prohibitive. Processes like precision investment casting must be automated before they are practical for automotive purposes.

The main thrust of Huebner's comments to the press at that time (late 1973) was that mass production of a passenger-car turbine appeared to be still a decade away. He figured it this way: four years for final design, development, and tooling; four years for limited production and large-volume tooling; and two years to get the volume production fully under way.

But Huebner concluded the conference on a bright note, and a subsequent development embellished the situation further. He predicted a whopping 50 percent improvement in fuel mileage in the gas turbine engine—which has a much lower octane requirement than gasoline or diesel engines—due to a projected improvement in weight-to-horsepower ratio and

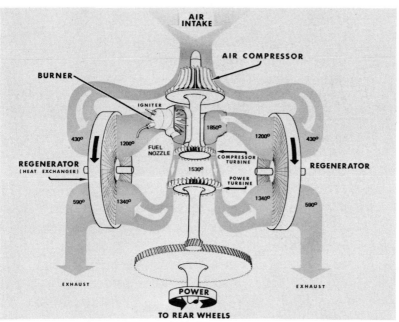

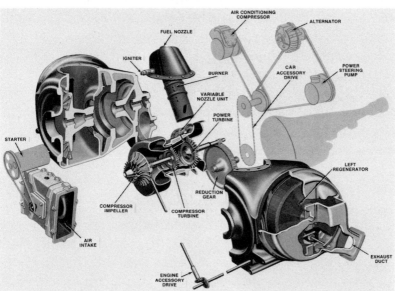

Exploded view (above) of Chrysler sixth-generation turbine engine designed for passenger-car use. Simplified schematic (below) of same engine. (COURTESY CHRYSLER CORP.)

to the probability of a higher operating temperature for the engine. He explained it this way:

> A 100-degree rise in temperature means a 6 percent improvement in fuel economy and a 14 percent improvement in specific output. Thus, a 500-degree rise—which we believe is attainable—means a 30 percent improvement in tank mileage.

Within a few months, in February 1974, scientists at the General Electric Research and Development Center, Schenectady, New York, announced a formidable breakthrough that will permit those higher temperatures that Huebner was talking about: the first simple and inexpensive technique for fabricating ceramic parts from silicon carbide, one of the most heat-resistant materials known. A compound of silica and carbon, two abundant and cheap substances, silicon carbide can potentially be fabricated into high-temperature turbine wheels at costs lower than those of the metal alloys presently required for gas turbine engines, which, incidentally, are in relatively short supply. The high-production requirements of any turbine designed for automobiles would dictate that common, not rare, materials be used.

Unquestionably, the gas turbine is beginning to be a success in trucks and other heavy-duty applications. Does that mean, then, that it will shortly be seen in passenger cars? If the history of the diesel is any indicator, the answer would have to be a resounding "No." But the histories of the diesel and the gas turbine are not parallel, and in this new age of pollution problems that directly block the solution of fuel-shortage problems, it's anybody's ball game.

The gas turbine is expensive to build but low in maintenance cost. That makes it sound like a good engine for heavy-duty commercial applications, where big businesses can afford to buy them and enjoy the trouble-free qualities. But the gas turbine has great pollution-free potential locked in with promising savings in fuel consumption. Those qualities could get the gas turbine into the passenger-car business in a hurry.

Outside view of Chrysler's sixth-generation turbine engine.
(COURTESY CHRYSLER CORP.)

7

All the Stirling Qualities

THERE is what is known historically as the "old" Stirling engine, and then there are the "new" Stirling engines. Almost 160 years separate old from new, and they are worlds apart in other ways.

The original was invented by the Reverend Robert Stirling, a Scottish clergyman, in 1816 as a pumping engine in mines. It used air at low pressure as its "working fluid," and it was manufactured in small quantities for many years, serving in a variety of applications where a simple, safe power source was required. But small steam engines, electric motors, gasoline and diesel engines came along, and beside them the old Stirling was inefficient and bulky. Today's engineers say that the Stirling became a victim of its competition due to the general lack of knowledge of thermodynamics a century and a half ago.

Modern technology made possible the revival of the Stirling by N. V. Philips' Gloeilampenfabrieken (Philips of the Netherlands) in 1938. At that time, Philips, a leading electronics manufacturer, was interested in developing a heat-driven electric generator for radios that could be operated in places around the world that lacked public utilities and where a common fuel would be easier to obtain and stock than batteries. For twenty years after that, except for interruptions due to World War II, Philips worked on the development of a new Stirling cycle engine that could compete in efficiency and output with internal-combustion engines. It had already become apparent to some scientists and engineers that the Stirling had qualities of quiet and vibrationless operation, use of a wide range of fuels, and a low-pollution exhaust that far exceeded internal-combustion engines and steam engines of both reciprocating and turbine types.

The Stirling is based on a closed cycle. If such an engine is internally pressurized with a gas and is heated at one place and cooled at another, it can drive a shaft, which will rotate and give mechanical power.

The Stirling is an externally-heated engine. In automotive applications heat can be supplied by a torch type of burner, which would provide external continuous combustion. That type of combustion, which occurs at near-atmospheric pressure rather than under compression inside an internal-combustion setup, is much more easily controlled, can be made more complete, and is therefore more pollution-free.

The Stirling has positive-displacement piston compression and expansion. It utilizes a sealed high-pressure working fluid (hydrogen or, with less efficiency, helium) and it operates at relatively low speed, providing great durability. However, it requires high-temperature alloys for its heater head at the combustor.

Philips has licensed other companies interested in proceeding with Stirling development. General Motors Research Laboratories, which for about ten years had been conducting an extensive study of external-combustion engines (including both vapor- and gas-cycle types), made a formal agreement with Philips in 1958 that initiated an intensive cooperative research and development program on Stirlings. GM was the major licensee during the 1960s, accumulating more than 25,000 hours of operating time on experimental Stirlings ranging from 3-hp single-cylinder powerplants to 400-hp 4-cylinder engines. GM allowed its Stirling license arrangement with Philips to lapse in 1970, and GM subsequently became more intimately involved with its new Wankel development program.

GM never attempted to develop an automotive Stirling engine. The GM program, instituted at Allison Division shortly after the launching of Sputnik in October 1957, had a space engine as its target. (GM did have an Opel with a hybrid Stirling-electric powerplant, but this was later revealed to be nothing more than a feasibility study of the mammoth research facility.) A number of years later, NASA's requirements for tiny sources of power were being adequately filled by the new fuel cell technology, and it was at this point, in 1970, that GM decided to drop the Stirling.

GM research engineers demonstrated that the Stirling has superior qualities of high efficiency and output, with efficiencies that are equal to or better than those of the best internal-combustion engines and about twice the thermodynamic efficiency of steam or organic-vapor engines. In large engines of several hundred horsepower, the Stirling is about the size and weight of a comparable diesel, in other words, big and heavy. In smaller sizes, 10- to 100-hp, the Stirling will tend to be somewhat larger than a gasoline engine, but it will be more efficient, according to GM's findings.

The Stirling has been found to be the quietest of all engines with automotive potential. A 3,000-watt engine-generator set built for the Army by GM is virtually inaudible at a distance of a hundred yards. And when fueled by diesel oil the Stirling had exhaust emissions far below the limits set by federal standards. Because it is an external-combustion engine, it has great fuel versatility—it can burn anything that can be provided through a torch. In fact, the Stirling can utilize any high-temperature heat source, such as stored heat, radioisotope heat, or some other nonburning source, such as chemical heat or radiation from an electric-element heater. Combustionless heat would of course operate the engine with zero exhaust emissions and, independent of the usual supply sources, the Stirling has special application possibilities in space or underwater situations.

Philips has also licensed United Stirling in Sweden and MAN (Maschinenfabrik Augsburg-Nürnberg) in Germany. United Stirling has been a major contributor to recent progress, particularly on heavy-duty—but not excluding passenger-car—applications. Both Philips and United Stirling have made prototype bus installations since 1970.

In August 1972, Ford Motor Company, Philips of Eindhoven, Netherlands, and U.S. Philips Corporation of New York City announced an agreement for a Stirling engine license and development program. The terms stated that Ford would obtain an exclusive world-wide license under Philips know-how and patents for car, truck, tractor, bus, military vehicle, industrial, and surface-vessel Stirling engines, and a nonexclusive license for all other Stirling engines. Ford obtained the right to sublicense to other companies on reasonable terms; and if Ford manufactures Stirling engines for the market, it will pay royalties to Philips. A seven-year development program was commenced.

Philips, one of the world's largest manufacturing corporations, is a major producer of electrical and electronic equipment. It initially became interested in the Stirling engine as a powerplant for electrical generator sets and a variety of applications, including torpedo propulsion, space power, and engines for boats and submarines.

Prior to the agreement with Ford, Philips had been worrying with the weight and bulk of the Stirling, especially where intended for automotive applications. In recent years the company had also been fighting excessive NOx emissions from its models. By the time of the Philips-Ford get-together, however, it was generally understood by engineers of both companies that solutions were possible for all those problems. Since Philips, Philips-MAN, Philips-United Stirling, and Philips-GM had already done considerable work with smaller engines with encouraging results, the new Philips-Ford venture proceeded directly into investigating the technical aspects of building a Stirling substitute for the Ford 351 cubic inch conventional engine then being installed in the intermediate-size Torino passenger car.

A Dutch engineer, Dr. Roelof Jan Meijer, who has been an employe of Philips since 1948, unquestionably has done more research and development work on the Stirling than anyone else. He sees a great future for the Stirling because it is nonpolluting and can comply with the postponed 1976 emissions standards whenever they are imposed; because of its low fuel consumption rate and its multi-fuel capabilities, both qualities to become increasingly important; and because it is a very quiet engine, and anti-noise standards are inevitable.

The fundamental principles of the Stirling are similar to those of the steam and (organic) vapor engines in that combustion is external and there is expansion of a working fluid within a cylinder containing a piston. A typical internal-combustion engine produces power by means of an episode of expansion of a compressed and heated volume of air (gas). This happens in the Stirling also, except that the heat is supplied from the outside. Whereas in the conventional engine, heat is generated by burning fuel inside the chamber where the expansion occurs, in the Stirling the heat is supplied by means of an external flame or other external heat source through a heat exchanger, called a heater head, to the working gas that is sealed inside the engine. The working gas is merely a medium for heat-energy transfer. Except for variations in temperature and consequent pressure, the gas undergoes no changes, is not consumed, and except in the case of an accidental leak, is not lost. Obviously, the sealing of the engine must be absolute, it must be leak-proof, and this is a major consideration in design and construction, and especially in forthcoming quantity production.

This is the primary, or simple, Stirling cycle: A cool volume of gas, entrapped by a piston, is first compressed and

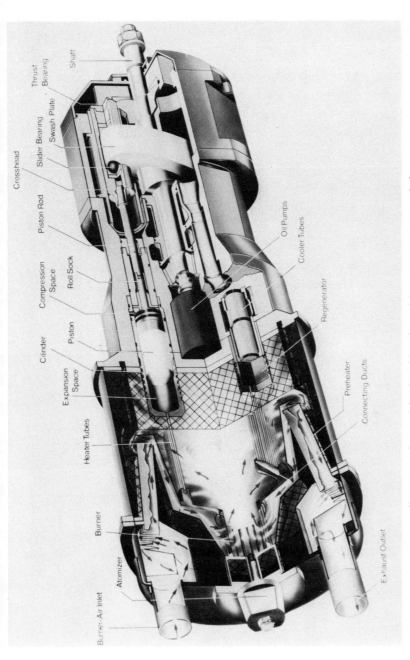

Cutaway view, with identification of important parts, of the Ford-Philips version of a modern Stirling engine. (COURTESY OF FORD MOTOR CO.)

Thrust Bearing

Shaft

Swash Plate

Slider Bearing

Crosshead

Piston Rod

Oil Pumps

Cooler Tubes

Compression Space

Roll Sock

Regenerator

Cylinder

Piston

Expansion Space

Preheater

Connecting Ducts

Heater Tubes

Burner

Exhaust Outlet

Atomizer

Burner-Air Inlet

then further heated by an external heat source. As the gas heats, its pressure increases, and the piston is driven downward to turn the crankshaft. After expansion, the gas is cooled by an external cooling source. Its pressure decreases, and the gas is once again compressed. Since the pressure that occurred during the hot expansion was much higher than the pressure present during the cool compression, there is a net work-output from the engine. The complete Stirling cycle takes place in one revolution of the crankshaft; in conventional four-stroke-cycle engines, two revolutions are required.

Reverend Stirling invented a refinement to overcome the cumbersome process of exchanging the heating and cooling sources. His invention replaced the alternating use of hot and cold sources by adding a mechanism, called a displacer piston, that served to move the gas between a stationary hot chamber and a likewise stationary cold chamber. The displacer is often referred to by engineers because it is a convenience in explaining the Stirling principle.

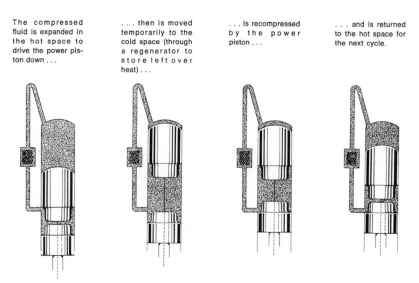

The compressed fluid is expanded in the hot space to drive the power piston down . . .

. . . then is moved temporarily to the cold space (through a regenerator to store leftover heat) . . .

. . . is recompressed by the power piston . . .

. . . and is returned to the hot space for the next cycle.

Simple Stirling cycle. The Stirling engine uses a working fluid—usually hydrogen or helium—that is contained in the engine. (COURTESY OF GENERAL MOTORS CORP.)

The compressed fluid is expanded in the "hot space" to drive the power piston down. Then the fluid is moved by the displacer piston temporarily to the "cold space" (through a regenerator, which is used to store leftover heat). Then the working fluid is recompressed by the power piston and is returned to the hot space for the next cycle.

However, a different arrangement, the modern double-acting piston, is used in modern technology in what is now known as the double-acting engine. Being smaller and less complex, the double-acting design is considered to be more suitable for automotive use.

Philips first constructed an engine with four separate but interconnected cylinders containing one piston each. The pistons were connected to each other by means of a device that kept them phased at 90-degree (quarter-circle) intervals. The phasing and interconnection caused each piston to act both as a power piston, in turn, and also as a displacer piston for the adjacent cylinder.

In addition to the interaction among the cylinders, the Stirling designed by Philips departed drastically from conventional engine configurations in two other ways: Rather than being in an in-line, vee, flat-opposed, or radial arrangement, the four cylinders are bunched together inside a single cylindrically shaped package, and the four separate sources of heat (burners) can be combined. Rather than being connected by rods and journals to a common crankshaft in the conventional way, the pistons bear through rods tipped with small wheels upon a device called a swashplate.

The swashplate is an inclined circular plate on a rotating shaft. It is capable of transferring, to the shaft, force and motion from the piston rods—which are reciprocating in a direction parallel to the axis of shaft rotation. The 4-cylinder swashplate engine of Philips is perfectly balanced and has four torque impulses per revolution, in the manner of an in-line eight or a V-8; but the power of an impulse in the Stirling by Philips is much less.

George J. Huebner, Jr., director of engineering research for Chrysler Corporation, remarked on some of the Stirling's negative aspects when he spoke to the International Motor Press Association at New York in late 1973:

> NOx emissions require additional control measures, since its combustion is similar to the gas turbine. It uses explosive hydrogen as a working gas at high pressure. Well over 200 brazings per cylinder (for heat exchanger, tube ends, etc.) require space-science technology. Stainless steel and expensive super-alloys have to be used for cylinder head and heat exchanger tubes. Mass production technology is not yet developed. Size is about 1.5 times that of the diesel engine. Load change is slow. It requires working gas pressure changes in the system, so vehicle driveability and performance are questionable.

In a 1973 report on a world-wide survey of future automotive powerplants, the Eaton Corporation, a Cleveland manufacturer of heavy-duty axles and other commercial automotive equipment, gave the Stirling the highest marks of all the types of engines it investigated. In its analysis the company commented on the extreme smoothness and perfect balance of the Stirling and pointed to its "very minor cyclical variations in torque." It cited the favorable torque curve that provided substantial torque increase with decreasing speed. It said that the Stirling appears to have the lowest emissions of all known engines and that the 1976 standards could be complied with at little penalty in fuel consumption or cost.

The report mentioned some control problems and the requirement for high-temperature alloys in the heater head as inserting a cost disadvantage, but these obstacles were dismissed as "not as formidable as once believed." It is important to point out, however, that Dr. Meijer, who is the world's top authority on the Stirling, has said that its hardware is "enormously expensive." It is only fair to state that this is not the opinion of a cost accountant. In his important supervisory position at Philips Research Laboratories, Dr. Meijer has not been much concerned with production matters. In fact, he has probably never made an all-inclusive cost study on the Stirling. It was his job to make the engine run smoothly and economically. His employer, Philips, is not an engine manufacturer. Ford, on the other hand, began cost studies immediately, but for competitive reasons probably will not rush the results into print.

While cost, yet to be determined by Ford, may or may not be a problem, the Eaton report indicated that size was not a difficult problem. Weight was described as competitive, noise level as very low, maintenance requirements as low, fuel consumption potential as "lower than any other contender" and across the broadest range of fuels, and durability was described as extremely high.

If all technical problems can be quickly solved, Ford has said that it will proceed with Philips to build a prototype 175-hp engine for installation in an intermediate like the Torino, build additional engines and expand testing to more prototype automobiles, and then, it is hoped that by 1980 or thereabouts, build a factory and tool it up for quantity production of Stirling engines.

The Stirling is an engine that possesses a number of superb qualities. And many of those qualities either never could be or no longer can be ascribed to today's conventional gasoline powerplant. The conventional engine is "establishment," however, in every sense of the word, and it remains to be seen whether in the next decade or so it can be dis-established by an engine as radically different, albeit superior in theory, as the Stirling.

8

The Piston Gets a Stratified Charge

FOR A long time, at least since before World War I, engineers have recognized that a spark-fired internal-combustion engine would perform better if extremely delicate control over its diet could be maintained and if at least two separate recipes could be supplied during the fuel-intake episode. Defenders of the conventional engine declared that it would be a better citizen, pollute less and waste less fuel, if it could be provided with a sometimes rich, sometimes lean, but always correct mixture of gasoline and air on demand. Sounds simple enough.

With all the incredibly complicated things that have been done to and for the automobile engine during the past seventy-five years, spraying the proper fuel mixture into it doesn't really sound like too tall an order. Indeed, it probably didn't sound terribly difficult even when what we now call the stratified charge system was first attempted about sixty years ago. A stratified charge system is a method of feeding an internal-combustion engine some form of heterogeneous fuel-air mixture that is changeable, so that the engine is always being fed the correct recipe for its need of the moment. Since Ricardo in England first did his stratified-charge work during World War I, many inventors and development companies have assembled various designs with middling success. Technical papers were presented as early as 1922; General Motors received a patent in 1926; and the Russians, in an attempt to improve fuel economy, built a stratified-charge engine in 1954.

Early in 1973, Honda of Japan had a stratified-charge engine that did not require unleaded gasoline and could meet the U.S. emissions standards for 1975 without a catalytic converter or a thermal reactor. Moreover, Honda had successfully converted a Chevrolet Vega, and General Motors was interested enough to send its president, Edward N. Cole, to Tokyo to meet with Soichiro Honda, who heads the motorcycle and auto concern.

"I went through the whole system with Mr. Honda," Cole related later. "I had the pieces in my hand. I drove two of their cars which had this equipment, plus the Vega they had converted. These were good cars. They have done a good job."

Yet in spite of the fact that GM, along with many other car manufacturers, had a limited-confidential-disclosure agreement with Honda, it decided not to pay the price Honda asked for a full-disclosure arrangement. The price, based partly on the size of the purchasing company, was very high for GM. Ford and Toyota did, however, pay the fees asked of them and are working on a full-disclosure basis. Lee Iacocca, president of Ford, said of Honda in 1973: "They have a new invention, one that we didn't think of." Honda has promised to put its new invention on 500,000 cars during the 1975 model year, about one-half of which will be exported to the United States.

The Honda method of supplying a stratified fuel charge to an engine is called CVCC, for compound vortex controlled combustion. In the new Honda engine, a standard piston-in-cylinder block is used but the cylinder head is modified by adding an auxiliary combustion chamber and an extra intake valve. The engine also has two carburetors instead of the usual single unit mounted on the intake manifold. One carburetor supplies a rich air-fuel mixture through the extra intake valve to the small auxiliary combustion chamber. The other carburetor provides an extremely lean mixture (little gasoline,

145

much air) to the main combustion chamber; the lean mixture is so lean that a spark plug couldn't ignite it and it would misfire. But the spark plug, which is located adjacent to the auxiliary chamber, easily ignites the rich mixture, the flame-front of which travels to the main chamber and fires the lean mixture. The average of the small-rich and the large-lean mixtures is very lean compared to the mixture supplied to a conventional engine.

The super-lean Honda mixture tends to burn slowly with more complete combustion. This sharply reduces the unburned hydrocarbons and the carbon monoxide in the exhaust; and it also cuts back significantly on oxides of nitrogen (NOx) because the controlled combustion occurs at lower temperatures, and formation of NOx is worse at higher temperatures.

The Environmental Protection Agency tested three CVCC Hondas at its Ann Arbor, Michigan, laboratory. The EPA reported that the engines were very responsive, acceleration was strong, and expressway speeds were maintained with adequate passing power in reserve. There appeared to be no significant fuel-economy penalty, and all three vehicles easily met the pollution standards set for 1975 and the interim 1976 standards also. But not for 1977, and that could be a serious problem for Honda—in order to meet the 1977 NOx standard, Honda would have to recirculate the exhaust fumes for further combustion. That would seriously degrade the driveability of

	1971	1972	1973	1974	1975	1976
HC	4.1	3.0	3.0	3.0	0.41	0.41
CO	34.0	28.0	28.0	28.0	3.4	3.4
NOx	—	—	3.1	3.1	3.1	0.4

United States automotive emission standards (stated in grams per mile of hydrocarbons, carbon monoxide, and oxides of nitrogen). The year-by-year schedule is subject to revision, postponement, or other modification of the EPA. (COURTESY OF ENVIRONMENTAL PROTECTION AGENCY)

the engine, and preliminary tests indicate there also would be an 18 percent drop in gas mileage. Even if the standards remain relaxed for a few years because of the fuel shortage, they will have to be met eventually.

With more than 4,000 hours of engine test time on both Vega 4-cylinder and Impala 8-cylinder stratified-charge engines of its own design, General Motors is still worried about that 1977 emission standard for NOx (a maximum production of 0.4 gram per mile). In 1974 the company declared that if it is faced with that standard, continued development of the stratified-charge engine "cannot be justified." The company added, however, that if the 1976 standards, which are somewhat easier to meet, are continued, "the stratified-charge engine is a viable production candidate." The stratified-charge principle, it should be noted, is applicable to the Wankel rotary as well as to the reciprocating engine, but rotary application would entail a prodigious amount of innovative engineering.

Incidentally, it was revealed in May 1974, by *Automotive News*, the prestigious weekly industry newspaper, that GM probably already had access to Honda's CVCC through a Japanese licensee, Isuzu, the company that makes the LUV minitruck for GM. GM is a major stockholder of Isuzu and shares technical information with the firm. A Honda official confirmed at a press conference that Isuzu was not enjoined from sharing CVCC information with GM and that Isuzu was not prohibited from selling a CVCC version of the LUV in the United States.

For a number of years, Ford worked with a fairly standard concept of stratified-charge engine fueling, primarily aimed at the goal of improving fuel economy, with results consistent with maximum power equal to that of a carbureted engine. That stage is remembered by Ford employes as the company's first-generation work in the Ford Combustion Process (FCP). Emission characteristics of the first-generation engine were far short of long-range objectives. Development in recent years has been aimed at the classic target:

147

minimizing exhaust emissions while maintaining maximum fuel economy.

Now in what the company identifies as its second-generation endeavors toward those often diametrically opposed goals, Ford is actively in pursuit of exhaust emission control by means of the Ford Programmed Combustion Process (PROCO). In Ford's newest process, emission reduction has resulted primarily from the use of air throttling and exhaust-gas recirculation, plus slight modification of other combustion-control parameters. PROCO remains basically a stratified combustion process, but it is so different from the standard aspects of stratified-charge combustion processes that the company chose to rename it; hence, PROCO.

PROCO includes as its foremost characteristic a special intake port that is shaped to impart a swirl to the fuel charge that describes a circle around the cylinder-bore axis. The swirl speed in the engine, in revolutions per minute, varies from three to five times the crankshaft speed, a rate that is much higher than in conventional engines.

Intake air is throttled at part-load so that the air-fuel ratio of the intake charge is maintained at 15.5 to 1—extremely lean—and 12 to 18 percent of the exhaust gas is recirculated into the intake system. The swirling fuel charge is compressed at a ratio of 11 to 1. The charge is compressed into a cup-shaped chamber in the piston head having about 65 percent squish area (the area under pressure during compression).

Fuel is injected at a point during the compression stroke by means of an injector that emits at a point near the center of the cylinder bore. The injector provides a soft, low-penetrating, wide-angle, conical spray that results in a rich mixture at the center surrounded by a leaner mixture and excess air. The spark plug, which may be located with its gap near the cylinder-bore centerline or just above the main spray, ignites the charge near top dead center.

Combustion begins in the rich mixture, progresses rapidly

The Ford PROCO engine mounted on an engine stand. (COURTESY OF FORD MOTOR CO.)

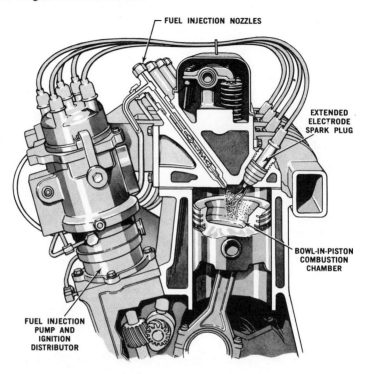

Ford's stratified charge internal-combustion engine—
PROCO. (COURTESY OF FORD MOTOR CO.)

through it and spreads out into the leaner regions. Thus it can be seen that Ford's PROCO system achieves many of the objectives of the stratified-charge principle by carefully aiming the injection of the hydrocarbon and precisely controlling the admittance of the air to be mixed with it, so that the entire process follows a planned program.

The several variations of stratified-charge systems being developed by major automobile manufacturers are the most important, but by no means the only, methods of purifying the piston. Other methods include lean-mix systems for improving fuel mileage as well as lowering emissions; heat pipes for transferring a certain amount of exhaust heat to the intake manifold for better vaporization of the fuel mixture; development of exhaust-treatment catalytic systems that are more

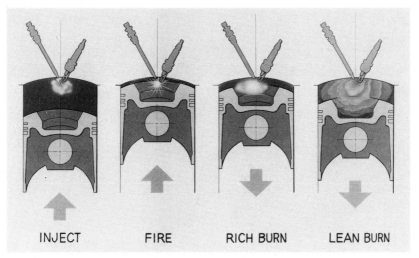

INJECT   FIRE   RICH BURN   LEAN BURN

In this artist's conception of a Ford PROCO engine, two portions of two strokes (compression, combustion) are portrayed, in which fuel is injected and ignited in such a way as to cause a small "rich burn" followed by a larger, longer "lean burn." (COURTESY OF FORD MOTOR CO.)

lead-tolerant; development of exhaust system filters for trapping lead particulates; and an interesting method of utilizing hydrogen as a supplemental fuel to be mixed with the gasoline-air charge in a conventionally carbureted system.

Under development in 1974 at NASA's Jet Propulsion Laboratory, Pasadena, California, the partial hydrogen injection method for internal-combustion engines promises an increase in efficiency and a decrease in emissions. The laboratory feels that the system has the potential of meeting even the 1977 EPA emissions standards with fuel economy superior to uncontrolled engines and using current fuels without loss of conventional response characteristics.

The system uses an on-board hydrogen generator that converts gasoline, water, and air to hydrogen and carbon monoxide. The generated gas is then mixed with gasoline and fed to a conventional engine. The relatively small amount of hydrogen entrained into the gasoline allows burning of the fuel at ultra-lean conditions. Because the generator produces all the hydrogen required, no hydrogen is stored aboard the vehicle.

Any successful modification of the existing spark-fired, reciprocating-piston engine that will increase fuel mileage and decrease exhaust emissions has the jump on all other engine developments—a multi-billion-dollar jump. Detroit has been accused repeatedly of dragging its feet in matters of "revolutionizing" the under-hood area of its automobiles. The billions tied up in tools for producing the kinds of engines we've always used are a formidable factor in any economic decision relating to a drastic change.

The stratified charge—providing the combustion chamber with a special mixture that is ideal for the moment—is a method of retaining the traditional piston engine for years and years longer. Many buyers say, "It's a good engine—why not keep it?" When provisions for keeping it exist—and especially if no one has come up with a superior competitor—the question is moot.

9

Alternatives to the Alternatives

THERE'S an automobile in Ada, Oklahoma, that's powered by a windmill. And in Pennsylvania a car is propelled by compressed air, while a Missouri auto that is otherwise conventional is hybridized with the addition of an exhaust-drive turbine engine mounted on its rear axle.

Automobiles have been powered by coal dust, gunpowder, wood, charcoal, coke, kerosene, diesel oil, alcohol, elemental hydrogen, garbage, manure, and uncounted other materials. It is possible to run a car on air and water, sunlight, and the inertia stored in a flywheel. The many machines man has come up with to utilize the plethora of fuel materials should be, in total, staggering; but apparently nothing staggers the imagination of mankind.

The Mother Earth people of publishing fame have successfully run a standard automobile engine on methane gas obtained as a result of natural biological decomposition of barnyard manure. To carry enough of that type of fuel—in anything short of a tagalong blimp—will, however, require the refinement of a few technological processes.

The windmill car, designed by Arnold R. Allen, is a battery-operated vehicle that can be parked in a blustery location and recharged via a collapsible windmill-generator unit set up on the car's roof. In May of 1974, Allen was having trouble obtaining suitable drive motors for cars of conventional size, so he was concentrating on the development of a much smaller three-wheeled vehicle that he considers suitable only for in-city operation. In mid-1974 Allen had made no plans for going into commercial production. Although subject to the usual limitations of all electric vehicles, Allen's car and its unique charging system are designed on principles that are basically sound. In the wide-open spaces where wind is frequent and free, Allen's idea has strong commercial possibilities. Several years earlier, Karl Bergey, a University of Oklahoma professor, designed a wind-powered generator for household use that had an annual output of 159,000 kilowatt hours at a cost of only two cents per kilowatt hour. In 1972 Professor Bergey's students were working on a much smaller windmill-generator set designed to recharge the batteries of an electric car.

The Pennsylvania air car gets its "charge" from an electrically-driven air pump that builds up a tankful of compressed air, which is used to propel the vehicle by means of air motors. Designed and owned by W. "Bill" Truitt of McKees Rocks, who formerly built racing cars in West Virginia and Ohio, the present-day air car is a development of Truitt designs that go back to the 1920s. The car Truitt was displaying in 1974, the "Pneumatic Electric Air Car," had been road-tested on the streets of McKees Rocks for more than eight months. The car is of compact size and has a heavy-duty plastic body with three interchangeable roofs that can convert the car from a sports model to a station wagon to a conventional sedan configuration. Truitt claims a maximum speed of 50 mph with the powerplant, a two cylinder V-type engine powered by compressed air from three tanks, one of one thousand pounds per square inch and two of two thousand pounds per square inch maximum pressure. The car has instant-start capabilities, and the engine runs almost silently, with its only noise, an exhaust hiss, dampened by a special muffler.

The Missouri invention, a brainchild of John Self of

Warrensburg, needs no muffler because its exhaust pipe from the manifold is connected directly to a small turbine engine that is installed on the left rear axle. The inventor, who is the operator of an ambulance service, calls his invention a "rotary muffler antipollution device." He claims that the exhaust pressure directed to the turbine adds sixteen to eighteen horsepower that otherwise would be simply vented to the atmosphere. He also claims that the car on which his turbine is installed, a ten-year-old Ford, gets an average of seven more miles per gallon than when it had only one engine, that the turbine acts effectively as a muffler, and that the turbine blades tend to "beat some of the emissions out" of the exhaust gases. It is important to note, however, that Self has not had any laboratory tests done on the emissions from his invention.

An inventor in Adelaide, Australia, has completed preliminary design work and built a prototype of a spherical engine he says will be "the smoothest-running and among the quietest internal-combustion engines ever built, with a life three times that of conventional engines but with only about one-half their weight." Called a "wobbling sphere" type, the Australian development has seven pistons around its equator that are shaped on one side to mate with the sphere's curve. As a fuel charge is ignited in a cylinder, the resulting thrust of the piston induces a nutational or wobbling type of oscillation in the sphere, and the wobbling motion revolves the transmission.

Cosworth Engineering of England worked for several years with the Chevrolet Division of General Motors on a "super engine" now known generally as the Cosworth Vega. The low-emission, high-performance powerplant, first offered in a limited production run in late 1973, is equipped with electronic fuel injection, breakerless electronic ignition, and an aluminum cylinder head with twin overhead camshafts and four valves per cylinder. The overhead camshafts are mounted in a separate carrier atop the aluminum head. One camshaft controls the intake valves and the other controls the

The Cosworth Vega engine mounted on a stand for display. (COURTESY OF CHEVROLET MOTOR DIVISION, GENERAL MOTORS)

exhaust valves, with a pair of each being assigned to each cylinder.

Each set of valves lies in planes inclined at 20 degrees from the centerline of the cylinder bores, providing a 40-degree included angle between intakes and exhausts. This forms "pent-roof" combustion chambers with a spark plug located at the peak of each. Four valves per chamber are provided to achieve the greatest possible valve area for a given chamber size. The intake valves produce high turbulence in the combustion chambers for more complete burning, higher efficiency, and lower emissions.

The electronic fuel injection system is a Bendix design to handle all combinations of manifold pressure and engine speed. A computer in the system provides continuous monitoring of manifold air temperature, manifold pressure, throttle-body position, and engine speed.

156

Cutaway view of the Cosworth Vega engine. (COURTESY OF
CHEVROLET MOTOR DIVISION, GENERAL MOTORS)

Data fed to the electronic control unit is assimilated into a signal sent to individual fuel nozzles in each port in the engine's aluminum intake manifold. The nozzles receive their fuel from an electric fuel pump at a constant 40 lbs/sq. in. The metering is so accurate, according to Chevrolet engineers, that only the minimum amount of fuel necessary to meet the engine's demand is consumed for each specific operating condition. The precise metering distributes fuel equally among the four cylinders, and good fuel economy as well as driveability result from the super-lean operation.

The standard Vega is rated at 72 hp, yet the Cosworth Vega with the same block develops 135 net hp at 6,000 rpm, with fuel economy in the 20 to 22 mpg area. This represents a rare attempt, so far, to aim a fuel-economy engine at a high-performance market, where it will be expected to mingle with imports like the Datsun 240Z, the BMW 2002, and the Alfa GTV. The expensive engine is being produced at the Tonawanda, New York, plant of Chevrolet.

The publishers of *Mother Earth News* built an experimental methane generator on Richard Shuttleworth's cattle farm near Redkey, Indiana. It has received a great deal of favorable publicity because it proved the practicality of making an engine fuel from nothing more than cow manure and water. An assistant to Shuttleworth, L. John Fry, produced methane more than twenty years ago on a thousand-head hog farm near Johannesburg, South Africa, primarily for the purpose of disposing of animal waste. Fry's "sump digester" converted organic wastes (plant, animal, or human) into methane—which is a useful gas that can fuel an automotive-type engine—and a nitrogen-rich fertilizer.

For many years, in the United States and elsewhere, most large municipal sewage disposal plants have had in use equipment for the recovery and storage of methane, which is a product of decomposing sludge. The valuable methanol gas is usually used only within the sewage treatment plant, for fueling large engines for pumping, for driving air condition-

ing equipment, and for heating purposes. No "methane industry" exists, however, and it is not reasonable to expect that in
the foreseeable future we will be operating automobile engines
on methane.

In 1973, the Environmental Protection Agency initiated
identical contracts with the Institute of Gas Technology,
Chicago, and ESSO Research and Engineering of New Jersey.
The objective of the contracts was to identify the most promising domestic non-petroleum derived fuels that would at the
same time permit a reduction in the automotive use of petroleum and provide acceptable performance, emissions, and
economy.

The initial list of targets set up for an investigation that
will require years is a formidable one. It includes checking on
the possibility of extracting gasoline from coal; a similar
distillate from shale; methanol (similar to wood alcohol) from
coal; ethanol (grain alcohol) from various carbohydrate
sources; hydrogen from coal by gasification and from water
by electrolysis and by thermal cracking; methane ("swamp
gas") from coal by gasification; ammonia from coal by gasification and from water by electrolysis; hydrazine (an oily,
fuming liquid, H_2NNH_2, known as a jet-propulsion fuel) from
ammonia by synthesis.

In mid-1974 it was the opinion of the contractors that
none of the fuels under consideration could be made available
in substantial quantities before the mid-1980s, except that in
the cases of fuels derived from coal existing technology with
definite shortcomings could be utilized to bring those fuels
on-stream a few years sooner while an improved technology
was being commercialized.

As the fuel situation becomes more interesting to re-
searchers, some Army scientists say they have developed a
process that could turn cow manure, table scraps, shrubbery,
trees, and grass clippings into a cheap auto fuel that can be
blended with gasoline in a passenger-car tank. The process
takes paper, trash, stable leavings, or whatever—anything

that's made of animal or plant fibers—and turns the stuff into a sugar called glucose, which can be fermented into a cheaply produced ethyl alcohol, or ethanol. The ethanol can be blended with gasoline to make a clean-burning fuel that will burn well in autos. The price, in production, is someplace between the 20-cent-per-gallon tag enthusiastically placed on it by Senator William Proxmire and the 35-cent figure that the scientists tend to favor—in any case, a bit cheaper than gasoline (before tax) and not as likely to climb out of sight as the oil companies continue to recover from their "profit shortage." The problem for now is that there is no alcohol industry capable of supplying our automotive needs.

There also is no hydrogen industry, and that's a real shame. Hydrogen, the elemental gas that is the chief fuel constitutent in gasoline and other fossil fuels, has two outstanding qualities that are woefully deficient in petroleum: it is abundant, virtually limitless and inexhaustible, because it can be produced from seawater and through coal gasification; and it is practically nonpolluting, burning cleanly and rapidly, with little besides water vapor as its exhaust product. There are some relatively minor problems to be overcome in converting from fossil fuels to hydrogen for automotive transportation: Hydrogen must be compressed to a liquid form, at which point it is 423 degrees below zero Fahrenheit and must be kept in a cryogenic storage tank. But that has been done without difficulty by the Billings Energy Research Corporation, Provo, Utah, using tanks developed for space use by Beech Aircraft Corporation. There is a certain explosion hazard, but scientists working intimately with hydrogen insist that it really is no more dangerous than gasoline, familiar to all Americans. Although hydrogen is available as a fuel, it is not yet produced in quantities large enough to make it economically competitive with conventional fuels.

One of the nation's first hydrogen cars was developed by Professor Robert Adt, Jr., at the University of Miami. As Lockheed Aircraft and NASA began in 1974 a joint study on the possibilities of hydrogen fuel for commercial jet aircraft, Professor Adt continued work with a Toyota he had converted to run on hydrogen. The Toyota's engine was rather simply set up to induct hydrogen directly into its cylinders. With no other modifications, the engine ran fifty percent more efficiently than it had been running on gasoline. The engine's exhaust contained only water vapor and a trace of nitrogen oxide (NOx).

"The changes we made on the Toyota," said Professor Adt, "were similar to the changes Detroit makes almost every year in manifolds and cylinder heads. The rest of the engine stayed the same."

Most scientists and engineers consider hydrogen to be theoretically an ideal fuel, and impatient cries for converting the automobile industry to hydrogen are being heard more and more frequently. We have a considerable way to go, however, before that attractive theoretical possibility becomes a reality.

Pure elemental hydrogen doesn't just lie around like a chunk of lead; it is always busy uniting itself with other elements to form compounds of various sorts. The earth's waters contain an unlimited hydrogen potential, but we don't have any low-cost way of extracting the element to use it as a fuel. Even though the price of petroleum rose remarkably beginning in 1973, oil is still a much cheaper fuel than hydrogen, and gasoline is still by far the cheapest form of portable energy. In a fair unit-of-energy comparison, with gasoline at sixty cents a gallon retail, hydrogen costs about three and a half times more per British thermal unit (BTU) than gasoline.

Even if we could afford to use hydrogen for fuel, it could not replace gasoline in five years, in ten years, and possibly not during this century, for a very basic reason: We were producing in the mid-1970s about 2.5 trillion cubic feet of hydrogen per year. At least a hundred times that much would be needed to sustain a hydrogen economy. It is not possible simply to expand the present plants or build enough new ones of the same kind to meet the demand of a hydrogen economy, according to

John E. Johnson of the Linde Division of Union Carbide, the world's largest producer of hydrogen. There are only two ways to obtain the needed massive supply of hydrogen, said Johnson in mid-1974: nuclear and solar plants for thermal cracking, and electrolysis extraction. Both methods will be used, Johnson predicted.

During the automobile's near-century of existence, it has been subjected to the inventive ministrations of man on many thousands of occasions. So many automotive ideas have emerged and reemerged that today we can hardly tell the new from the old. A rotary engine was introduced by Adams and Farwell in their innovative 1906 car, which also had a removable steering wheel and control pedals to convert it from a two-seat coupé to a single-seat roadster. The 1905 Duryea Phaeton had a single control for steering, accelerator, and gearshift similar to the equipment on a modern "experimental" car of a few years ago.

Double-burning—returning some exhaust gases to the engine for a second burning to reduce pollution—was a feature of the 1906 Compound, which also offered power brakes. The 1903 Knox offered an air-cooled engine that needed no antifreeze; the White Model H touring car of 1907 had shock-absorbing bumpers; and the Thomas Model 35 offered safety belts in 1907. Add the inventions since 1907 and the list seems endless. Men will continue to improve the automobile, and especially in these times when fuel and pollution problems threaten the very existence of the automotive industry, we can expect startling new ideas to come on the scene with increasing frequency.

Perhaps some old principle will be reevaluated and found appropriate for our present times; possibly only original new thinking will suffice. But changes will come, you can be sure, and they will be welcome, valuable, and occasionally vital.

10

Which ... and When?

ending   statement

WHATEVER course future automotive engine development may take, three factors will determine it: cost of the powerplant itself; how much fuel it uses; the nature of its exhaust emissions.

An incidental factor, one that has not had a powerful influence on our automobile picture heretofore, will take on increasing importance from now on: a continuing trend toward specialization in automobile usage. A basic change in attitude toward the automobile began for Americans shortly after World War II when automobiles ranging from small to tiny were brought over from Europe as novelties. After a slow start, during which the prophetic little Rambler achieved surprising popularity while Henry Ford II was still declaring that his company would never build a small car, the compact car gradually became popular in the United States. In the typical two-car family the little car was the "second car" used by the housewife for errands, while the larger car was used by her husband for business and was almost always employed as the family vacation car. But both cars had multi-purpose engines: either car could be used for errands or for high-speed inter-city traffic. Yet typically the smaller car handled the shorter-trip work.

The cars of tomorrow probably will be even more specialized. More than ever before, cars will be separated into urban and rural applications, for example. Their power plants will gradually become less multi-purpose.

If the conventional engine of the 1970s is replaced during the remainder of this century, the replacement will be very gradual and by several alternatives, not just one new engine that will take over and achieve the dominance of today's engine. Changes from now on will be many, rapid in comparison with the past, and, for the automobile buff at least, fun to watch.

In mid-1974, alternative-engine programs of investigation were being conducted by the Environmental Protection Agency, the Department of Transportation, the National Aeronautics and Space Administration, the Atomic Energy Commission, the Postal Service, and the National Science Foundation. Moreover, many private agencies not directly connected with the automotive industry, such as universities, testing and research organizations, and corporate investors of venture capital, continue to seek the best new method of automotive propulsion. Several state governments, California being outstanding among them, are spending tax money in the search. And—contrary to public opinion—the car makers themselves are forking over the really big money. The American companies alone spent more than $72 million in 1972, about $90 million in 1973, and even higher amounts were projected for 1974 and 1975.

In the early 1970s, pollution problems were at the forefront of automotive engine research. Management of exhaust emissions was a top-priority goal, and all car makers were mandated by government to reduce noxious products of combustion according to a set schedule, which became more stringent from year to year. The fuel problems of 1973 loosened the schedule somewhat because many antipollution measures wasted gasoline, and car manufacturers needed more time to perfect devices for lessening pollution without further losses of fuel economy.

One such device was the catalytic converter, for chemically changing most of the exhaust products into harmless nonpolluting substances. It was introduced on most American cars on the 1975 models that were first marketed in the fall of 1974. The "cat" converter, which requires unleaded gasoline, preempted its alternative, the thermal reactor, an afterburning device that wastes fuel because it requires a retarded spark and a rich mixture in the engine. Whether the industry acted too fast in adopting the catalytic method remains to be seen, but the National Academy of Sciences set off an alarm a full year before the converters were on the road. In 1973, Philip Handler, the Academy's president, wrote the following as part of an introduction to the report of the Committee on Motor Vehicle Emissions:

> The committee is concerned that mass production of what are presently deemed to be relatively fragile, catalyst-dependent systems, of unproved reliability in actual service, may engender an episode of considerable national turmoil. It is further concerned that, once committed to the manufacture of catalyst-dependent control systems, rather than switch to some more generally acceptable system such as a version of the stratified-charge engine that now offers great promise, the relatively ponderous automobile industry will continue to manufacture catalyst-dependent systems for some years. . . .

Whether the conventional reciprocating-piston engine remains dominant throughout the rest of the twentieth century, or whether one or several other designs take its place, depends on fuel availability and ability to control pollution, because no other engine of the foreseeable future can compete on an initial-cost basis with the conventional powerplant. Many engineers feel that the conventional engine cannot long survive the dual pollution-fuel bind, but among those who suggest alternatives there is wide disparity of opinion.

The prestigious 1973 report on a world-wide survey of future automotive power plants by the Eaton Corporation, a

Cleveland manufacturer of highway transport equipment, discounts electric vehicles and steam engines in assessing the promise of various engines to come in the near future. Yet limited quantities of electric vehicles are being built right now. And Bill Lear, who has been right enough of the time to provide the world with electronic marvels, one of the greatest of all jet airplanes, and hundreds of other innovations we use routinely, feels that the steam turbine is just over the horizon —a year or two away.

Chrysler and Ford heavily discount the Wankel. Henry Ford II said in 1974 that the little rotary was X-rated at his firm, and that "X means out, cross it out." Ford Motor Company is working hard on its programmed combustion process (PROCO) and is interested enough in Honda's CVCC (compound vortex controlled combustion) to pay a fat license fee for it. Ford also is working closely as a licensee with Philips to continue development of the Stirling cycle and perhaps perfect a Stirling engine for automotive use.

Chrysler, while not neglecting any of the other alternatives except the Wankel, is confidently betting on the gas turbine as the powerplant of tomorrow. General Motors, with many more millions invested in the Wankel than any other company has invested in anything, stated confidently in 1974:

> General Motors remains very much committed to the rotary engine. A great deal of work was necessary to bring the fuel economy to a level that we feel is satisfactory. Meeting the current and projected emissions requirements, while maintaining this fuel economy, has—as expected—posed a very difficult challenge. However, we feel the efforts required to resolve the problems associated with the rotary engine will reap significant rewards in terms of more efficient vehicles better attuned to our future needs. As indicated in the past, the inherent size and weight advantages of the rotary engine make it particularly attractive for vehicle packaging where space is at a premium. Therefore, the same advantages we saw in the engine initially not only remain but become more attractive in view of the market shift

to smaller vehicles. For these reasons, General Motors is still deeply involved in the preparation of the rotary engine for production release.

In assessing future engine possibilities, a diligent investigator will find almost as many opinions in Detroit as there are engineers. The enormous resources of the car manufacturers—a company might put four or five *thousand* people to work on a single project—sometimes seem to work against straight, logical thinking and following through on direct courses of action. Unfortunately, today's sophisticated marketing procedures often dictate that sales decisions impinge upon design decisions. Rarely can an individual make an important decision; and the favorite engineers' saying, "A camel is a horse that was designed by a committee," has a powerful ring of ironical truth to it.

Moving the short distance from Detroit to Ann Arbor can, however, uncover a fresh, non-proprietary point of view from a person who is very close to what this book is all about. He is John J. Brogan, Chief, Alternative Automotive Power Systems Division, Environmental Protection Agency. He gave this summary of his feelings in the fall of 1973:

> I expect that the intermittent combustion heat engine with combustion modifications and with exhaust aftertreatment will dominate through this current decade; thereafter, by the mid-eighties, at the latest, some continuous combustion heat engines not requiring exhaust aftertreatment may dominate new car engines. At the present time the gas turbine appears to be a leading contender. In the second half of the eighties, the all-electric may be competing with the continuous combustion heat engine for a very limited share of the automobile market.

In the middle of the 1970s, the Wankel engine stood out as the most controversial of all the powerplants that are serious contenders for supremacy in the automobile field. Some respected sales engineers forecast a possible Wankel penetration of as much as 25 percent into the market of the middle 1980s while others ascribe no more than 2 to 3 percent to the little rotary. But aside from the differences among the engineers over the Wankel rotary, nearly a million copies of which are already on the world market, there is fairly general agreement regarding the likely order of appearance of the other contenders. Barring unforeseen breakthroughs, this is the chronological ranking that has received the most favor:

The diesel. Rudolf Diesel's tough machine has the longest production history and the greatest production record next to the gasoline engine. The technology already exists for developing the diesel into a passenger-car type of package; and, short-term, it unquestionably is the best prospect for future perfection and marketing for automobiles.

Stratified charge. One or several versions will be on the market in quantity soon. Expense will be a drag on its popularity while the diesel temporarily gets a jump ahead; but well into the 1980s, stratified-charge modifications of the conventional gasoline reciprocating-piston engine should achieve a strong market position.

The gas turbine. Perfected for aircraft during World War II and for trucks in the middle 1970s, the gas turbine has already been close to the automobile market more than once. Two or three fortunate breakthroughs, said to be likely and imminent, could put a gas turbine for automobiles in production and on the market by the middle 1980s.

The steam (or vapor) engine. Rankine-cycle engines, either organic or water-based (probably both), and either reciprocating-piston or turbine (probably both) are almost a sure bet for sometime in the late 1980s.

The Stirling. Like its first cousin, steam, the Stirling will be standing in the wings awaiting a powerful demand, which could result from continued pollution and fuel-supply problems. But the Stirling, with its space-age technological requirements, is far down the road.

Electric-powered cars are the farthest down the road of all the alternative possibilities. Although, short-term, they will achieve a measure of popularity as urban and suburban errand-running vehicles, their prospects as highway cars will remain dim until they are provided with capable battery power. That will require a big breakthrough in battery technology.

"If we can put men on the moon . . ." has become a tiresome way of opening a discussion as to why ponderous old Detroit can't solve the pollution and fuel-shortage problems simultaneously. If an increase in income taxes were to make possible a throwing of billions of dollars into the quest for the "perfect" engine of tomorrow, the government would undoubtedly be able to come up with something as the result of another incredibly wasteful spending spree.

But that isn't the way to do it, and that's not the way it will be done. The engine(s) of tomorrow will be the results of free-enterprise research and development of yesterday and today. When the long term is compared with the short term, two facts stand out as certainties:

The engine of 1990 could be fantastically different from anything we have today.

The engine of 1980 has already been invented.

Bibliography

BOOKS

Ansdale, Richard F.: *The Wankel RC Engine*. A. S. Barnes & Co., South Brunswick & New York, first American edition, 1969.

Automobile Manufacturers Assn. Inc. (now Motor Vehicle Manufacturers Assn. of the U.S.) : *Automobiles of America*. Wayne State University Press, Detroit, 1962.

Crowe, Bernard J.: *Fuel Cells, A Survey*. (NASA), Computer Sciences Corporation, Falls Church, Virginia, 1973.

Dark, Harris Edward: *The Wankel Rotary Engine, Introduction and Guide*. Indiana University Press, Bloomington & London, 1974.

Donovan, Frank: *Wheels for a Nation*. Thomas Y. Crowell Co., New York, 1965.

Ford, Henry II: *The Human Environment and Business*. Weybright & Talley, New York, 1970.

Greenleaf, William: *Monopoly on Wheels*. Wayne State University Press, Detroit, 1961.

History of Research and Development on Mazda Rotary Engine. Toyo Kogyo Co. Ltd., Hiroshima, 1972.

Jamison, Andrew: *The Steam-Powered Automobile, An Answer to Air Pollution*. Indiana University Press, Bloomington & London, 1971.

L'Editrice Dell'Automobile LEA, Editor: *World Cars*. The Automobile Club of Italy & Herald Books, Bronxville, 1972.

Matteucci, Marco: *History of the Motor Car*. Crown Publishers, Inc., New York & Turin, 1970.

Nitske, W. Robert & Wilson, Charles Morrow: *Rudolf Diesel, Pioneer of the Age of Power*. University of Oklahoma Press, Norman, 1965.

170

Norbye, Jan P.: *The Wankel Engine, Design, Development, Applications.* Chilton Book Co., Philadelphia, New York, London, 1971.

Rotary Engines. Toyo Kogyo Co. Ltd., Hiroshima, 1971.

Stein, Ralph: *The Treasury of the Automobile.* Golden Press, New York, 1961.

Stern, Philip Van Doren: *A Pictorial History of the Automobile.* The Viking Press, New York, 1953.

Summary Report of the Advanced Automotive Power Systems Contractors Coordination Meeting, Ann Arbor, Michigan. Advanced Automotive Power Systems Development Division, U.S. Environmental Protection Agency, June 5–7, 1973.

Symposium on Low Pollution Power Systems Development, Ann Arbor, Michigan. The Office of Air and Water Programs, U.S. Environmental Protection Agency, Oct. 14–19, 1973.

ARTICLES

Agnew, Dr. Wm. G.: "Automotive Powerplant Research." General Motors Corporation, May 14, 1973.

Armagnac, Alden P.: "Flywheel Brakes." *Popular Science,* February, 1974.

"Auto Engines of the Future." *Changing Times,* March, 1974.

Bohn, Joseph J.: "SAE Reviews Electric Cars—They're Still Down the Road." *Automotive News,* March 25, 1974.

Burck, Charles G.: "A Car That May Reshape the Industry's Future." *Fortune,* July, 1972.

Cain, Charles C.: "None Rushing to Market Gas-Saver." *Associated Press,* January 16, 1974.

Callahan, Joseph M.: "Blowing Whistle on Steam." *Automotive News,* August 5, 1968.

Chaves, Robert: "Electrics: Cars of the Future?" *Automotive News,* March 18, 1974.

"Cracking the World Market" (editorial). *Automotive News,* May 27, 1974.

"Cummins Readies 'K' Engine Debut." *Automotive News,* October 1, 1973.

"Daily Use of Solar Energy Decades Away." *Associated Press,* February 13, 1974.

"DDA Stepping Up Output of Gas-Turbine Engines." *Automotive News,* October 1, 1973.

del Core, Sergio Favia: "Italians Develop Small Electrics." *Automotive News,* April 29, 1974.

Bibliography

"Double-Engined Car Made by Students." *Associated Press*, May 29, 1974.

Driscoll, James G.: "Exhaustive Job Pays Off; Little Honda Beats Big Makers to Low-Emission Engine." *The National Observer*, August 11, 1973.

"Electric Vans Rolling." *Automotive News*, February 4, 1974.

"EPA and Mazda to Join in Further Wankel Testing." *Automotive News*, March 4, 1974.

"EPA to Retest Mazda Against 3 Other Cars." *Automotive News*, March 11, 1974.

Ethridge, John: "50,000-Mile Mazda R-100 Rotary Tear-Down Report." *Road Test*, January, 1972.

Fanning, Patricia: "Charged-Up Car." *The National Observer*, July 13, 1974.

Farmer, Frank: "Will Electric Car Ease America's Fuel Crisis?" *Springfield (Mo.) Sunday News & Leader*, December 2, 1973.

Fendell, Bob: "Does GM Have Entry to CVCC via Isuzu?" *Automotive News*, May 27, 1974.

Fendell, Bob: "Why Chrysler Nixed Turbine Car in '67." *Automotive News*, November 19, 1973.

Fisher, Dan: "Bill Lear Stubbornly Clings to Steam." *Ward's Auto World*, May, 1972.

Hanley, Daniel Q.: "Dinnertime Leftovers May Run Your Car." *Associated Press*, April 27, 1974.

"Have Windmill, Will Travel." *Associated Press*, March 21, 1974.

"Hydrogen Car Making Progress." *Automotive News*, March 25, 1974.

"IH to Build Turbine Plant in San Diego." *Automotive News*, April 1, 1974.

"Introducing: GM's 1975 Rotary Combustion Engine." *Road Test*, September, 1973.

"Is the Wankel the Auto Engine of the Future?" *Changing Times*, July, 1972.

Kahn, Helen: "How Washington Views the Car of the Future." *Automotive News*, November 19, 1973.

Kidd, Stephen and Garr, Doug: "Can We Harness Pollution-Free Electric Power From Windmills?" *Popular Science*, November, 1972.

"Little Black Box Replaces Nose." *Automotive News*, February 25, 1974.

172

"Lower Emissions Reported in California Steam-Car Test." *Automotive News,* August 20, 1973.

"Mazda Disputes EPA Test Data." *Automotive News,* February 18, 1974.

"Mazda MPG Improves in EPA Retest." *Automotive News,* April 15, 1974.

Miller, William C.: "SPS Steam Car Ready for Tests." *Automotive News,* May 27, 1974.

Mitchell, Michael: "Steam Car Needs Work, Says Lear." *Associated Press,* February 6, 1974.

"New Angle for Electrics, A." *Automotive News,* June 3, 1974.

"New Combustion System Cuts Emissions—Perkins." *Automotive News,* June 17, 1974.

Norbye, Jan P. and Dunne, Jim: "Stirling-Cycle Engine Promises Low Emissions Without Add-Ons." *Popular Science,* February, 1973.

"NSU Wankel Tests at Thirty Six Miles a Gallon." *Automotive News,* January 28, 1974.

"Perkins to Build Diesels in America." *Automotive News,* March 25, 1974.

Pond, James B.: "OMC's Men and Their RC Machine." *Automotive Industries,* May 15, 1972.

Pope, Leroy: "Gas-Saving Items Find Ready Sale." *United Press International,* May 8, 1974.

Ragone, David V.: "Review of Battery Systems for Electrically Powered Vehicles." Paper No. 680453, May, 1968, *Society of Automotive Engineers.*

Riley, John William: "Trend Seen Favoring Steamers." *Automotive News,* July 30, 1973.

Roe, Jim: "First Report: PS Drives OMC's Hot New Stack-of-Wankels Outboard." *Popular Science,* June, 1973.

Rowand, Roger: "New Engines Draw Attention at GM's 'Show and Tell.'" *Automotive News,* February 18, 1974.

Saltus, Richard: "Hydrogen Car Gets Another Look." *Associated Press,* October 25, 1973.

Scott, David: "New Stirling-Powered Zero-Pollution Car Runs on Stored Heat." *Popular Science,* June, 1974.

Shuttleworth, John: "Mother's Methane Maker: Past, Present and Future." *The Mother Earth News,* Reprint No. 172.

"Silicon Carbide—Key to Turbine?" *Automotive News,* February 25, 1974.

Bibliography

Simko, A., Choma, M. A., and Repko, L. L.: "Exhaust Emission Control by the Ford Programmed Combustion Process—PROCO." Paper No. 720052, January 10, 1972, *Society of Automotive Engineers.*

"Sleek Electric-Powered Vehicle Undergoing Two-Year User Tests." *Fleet Management News,* January, 1974.

Smith, Dave: "Chrysler Tells WWR Why It's Lukewarm on Wankels." *Ward's Wankel Report,* August 25, 1972.

"Those Someday Sources of Energy." *Exxon, U.S.A.,* Third Quarter, 1973.

"Turboliner Set for In-Fleet Use." *Automotive News,* February 18, 1974.

"Two Fuel Storage Systems Debut on Hydrogen Car." *Automotive News,* March 25, 1974.

"VW Wankel Car to Give Buyers a Third Choice." *United Press International,* February 2, 1974.

Wadsworth, Nelson: "Plenty of Gas—If You Turn to Hydrogen, That Is." *The National Observer,* February 23, 1974.

"Wangle Yourself a Wankel." *Forbes,* December 15, 1972.

"Wankel Fuel Rap Called Misleading." *Automotive News,* March 25, 1974.

"What's a Wankel?" *Ward's Wankel Report,* July 13, 1973.

Williamson, Don: "Cole: Run Cars on Air and Water." *Automotive News,* September 24, 1973.

"Windmill Will Recharge Electric Cars." *Associated Press,* April 6, 1974.

Woron, Walt: " 'High-Mileage' (Tyce-Fish) Carburetor on Tap." *Automotive News,* May 27, 1974.

Woron, Walt: "McCulloch Builds Special Electric." *Automotive News,* March 25, 1974.

Wright, Richard A.: "Cosworth Vega—Hot, Clean." *Automotive News,* August 27, 1973.

Wright, Richard A.: "SAE Samples Technological Goodies." *Automotive News,* March 11, 1974.

Wright, Richard A.: "Stirling Prospects Good." *Automotive News,* September 24, 1973.

175

Index

Index

Index